the Milky Way galaxy

the Solar system

On the Cosmic Horizon

Ten Great Mysteries for Third Millennium Astronomy

Jeffrey Bennett

Addison
Wesley
Longman

Library of Congress Cataloging-in-Publication Data

Bennett, Jeffrey O.
   On the cosmic horizon : ten great mysteries for third millennium astronomy / Jeff Bennett.
     p. cm.
   Includes index.
   ISBN 0-321-02971-2
   1. Astronomy. 2. Astrophysics. 3. Cosmology. I. Title.

QB43.2 .B45 2000
52'.9'05—dc21   00.044182

2 3 4 5 6 7 8 9 10-BTP-05 04 03 02 01 00

This book is dedicated to Grant and all the other children who will grow up in the third millennium of the Gregorian calendar. May it be an age of enlightenment, free from the darkness that plagued millennia past.

Contents

Preface

Outside of religion, no human pursuit deals with deeper questions of our existence than astronomy. Thus it is no wonder that astronomy has long captured the public imagination. Stories of astronomical discovery often make the headlines, and astronomical images and ideas infuse many aspects of our culture. Today, however, the pace of astronomical discovery is so rapid that even professional astronomers sometimes have difficulty staying current. For everyone else, it's hard just to keep track of what professional astronomers are looking for, let alone to understand what they find. That's probably why friends and acquaintances who know me as an astronomer often ask, "What are the big questions in astronomy today?"

This book is my answer. Its primary purpose is to help you understand and enjoy the stories that are likely to dominate the astronomical headlines over the next decade or more. The book is designed to be accessible to anyone with an interest in the great mysteries of our universe, regardless of whether you've studied much astronomy before. As you'll see if you flip through the pages, it takes the form of a Top 10 List of cosmological questions that I call "mysteries for third millennium astronomy" to emphasize that they will be among the most active topics of research in coming decades. I've deliberately avoided questions that most astronomers consider settled, as well as questions that are unlikely to be answered in the near term (such as why the universe exists). In essence, the list is my personal guess as to which questions will appear most often in news reports of the early twenty-first century. There will undoubtedly be new and surprising discoveries that I have not anticipated and great progress in topics that I did not include in the top 10. And, of course, I apologize in advance to those astronomer friends whose favorite mystery didn't make my list.

In writing this book, I've tried to make the ten mysteries as independent and self-contained as possible, so you can skip around and read them in any order. At the same time, however, each of the mysteries introduces a few fundamental astronomical concepts, and I've tried to put these in an order that will enable you to build your understanding gradually. Thus, if you read the book straight through, by the end you'll have the beginnings of a solid background in modern astronomy.

So I invite you now to join in the human adventure of astronomical discovery. As you will see, we are on the verge not only of great new discoveries in science but of discoveries that may change the very way we look at ourselves as individuals and as a species. Happy reading.

Boulder, Colorado
August 2000

Introduction: The Mysteries Ahead

The universe is full of magical things patiently waiting
for our wits to grow sharper.

EDEN PHILLPOTTS (1862–1960)

Picture yourself near a mountain lake as twilight falls on a crisp, clear summer evening. You've arrived here on foot, with only the sounds of nature for company. Sitting near the water's edge, you watch patiently as the sky fills slowly with stars. Soon the band of the Milky Way comes into view, stretching high across the sky and shining with a brilliance city dwellers never see. The depths of the heavens are reflected in shimmering lights on the lake, creating an image of infinity both above and below you. If you set your mind free, you cannot help but ask some of the most fundamental questions imaginable. What is the universe and what is Earth's place in it? How did it all come to be? As you let your thoughts wander through time and space, you could be anyone at any time in history. After all, your questions are the same ones that humans of every culture have asked since history began.

But there's a difference. You are living at the start of the third millennium of the common era, at a time when the boundaries between the known and the unknown are shifting faster than they ever have before. Questions that seemed imponderable just a few decades ago are now all but settled, and many of today's biggest questions could not even have been imagined by our ancestors in centuries and millennia past. Of course, we still don't know why the universe exists in the first place or why we are here to ask such questions, and it's possible that we never will. But modern science offers satisfactory answers to many of the "what" and "how" questions that have puzzled humans for thousands of years. So before we move on to discuss the mysteries that lie ahead, it's worth looking back at the mysteries of previous millennia that already have been solved.

If anyone had bothered to make a list of the top ten mysteries of the universe at the dawn of the last millennium, it would undoubtedly have been dominated by questions concerning the layout of the heavens and the phenomena of the sky. What are stars and planets? How big and how far away are the Sun and Moon? Why do the planets and stars move through the sky as they do? And so on. Indeed, these questions probably would have been on the list at the start of the first millennium, and perhaps even on the list a thousand years before that.

The biggest mystery of all dealt with Earth's place in the universe. Until Kepler and Galileo settled the matter some four hundred years ago, Earth was generally assumed to be the center of the universe. But a few dissenting voices were heard through the centuries. In about 260 BC, the Greek astronomer Aristarchus reasoned that the Earth must go around the Sun, rather than vice versa. His contemporaries and later scholars rejected his arguments, but not without at least some impassioned debate. In 1440, Nicholas of Cusa wrote a book titled *De docta ignorantia* ("On Learned Ignorance"), in which he claimed that the Earth turns on its axis as it orbits the Sun, that stars are other suns and hold other inhabited worlds in their grasp, and that the heavens are infinite in extent. Interestingly, while Galileo was punished by the Church for holding similar beliefs two centuries later, Nicholas was ordained a priest in the same year his book was published, and later was elevated to cardinal. Thus we see that science and religion were not always in conflict, and there is no need for them to be in conflict today.

It was another Nicholas—Nicolaus Copernicus—who set in motion the wheels of science that would finally prove Earth to be a planet orbiting the Sun. We'll discuss this story in greater detail in Mystery 8, but for now we should look at the modern picture of the universe that has since emerged.

Today we know that Earth is the third (in order of distance) of nine planets that orbit the Sun. The Sun, the nine planets and their moons, and a myriad of smaller objects including asteroids and comets make up what we call our *solar system*—one of more than 100 billion star systems that make up the huge, swirling disk of the *Milky Way galaxy*. Our location in the Milky Way is a little over halfway from the galactic center to the edge of the galactic disk.

The Milky Way galaxy is one of the two largest galaxies in a group of thirty or so galaxies that we call the *Local Group*. (The other large member is the Great Galaxy in Andromeda.) Many other groups of galaxies are scattered throughout the heavens, and those groups that contain more than about a hundred galaxies are often called *clusters* of galaxies. The groups and clusters, in turn, often appear to make up larger structures called *superclusters*. Not surprisingly, the supercluster to which the Local Group belongs is called the *Local Supercluster*. Finally, the *universe* encompasses all the super-clusters and everything within and between them. Said differently, the universe is the sum total of all matter and energy. The post-card shown here summarizes the basic lev-els of structure in our cosmic address, and the painting on the inside front cover shows the same idea in a more visual way.

Questions about the origins of Earth and the universe might also have made the list of mysteries in past millennia, though many philosophers probably considered those questions unan-swerable by science. After all, until recently no one could imagine methods of studying the past such as the radioactive dating of rocks. Only a few brave souls even ventured guesses about how a planet might have developed. Still, one of the earliest guesses was uncannily modern; before 400 BC, the Greek philosopher Democritus

The Earth and Moon
One of the first photographs containing both the Earth and the Moon, obtained during the Galileo mission to Jupiter, when Galileo performed a fly-by of the Earth.
Photo by NASA

Megan!
You wouldn't believe what just happened to Me. We went sight-seeing on the third Moon of Creol and I swear that I saw your brother there. We tried to catch up to him but never found him. But the Moon was fabulous and so was New Faithful. Lots of love,
John

Megan Donahue
Space Telescope Science Institute
Baltimore, MD
USA
Earth
The Solar System
Milky Way Galaxy
The Local Group
The Local Supercluster
The Universe

This postcard shows how a distant friend might address a postcard to someone on Earth.

suggested that the universe began as a chaotic mix of atoms that slowly clumped together, ultimately forming the Earth and other worlds. Today's scientific story of creation similarly holds that simple components gradually developed into galaxies, stars, planets, and life. The difference between the modern story of creation and that envisioned by Democritus lies mainly in the details—and, more important, in that scientific evidence now makes this story far more than a guess.

According to modern science, the universe began somewhere between about 10 billion and 16 billion years ago with an event called the *Big Bang*. As we'll discuss in Mystery 4, several strong lines of evidence support the idea of a Big Bang. But the simplest evidence is the most direct. As first discovered in the 1920s by Edwin Hubble (for whom the Hubble Space Telescope is named), the universe is expanding in the sense that the average distance between galaxies is increasing with time. If intergalactic distances are growing with time, we can logically conclude that galaxies must have been closer together in the past. By running the current rate of expansion backward, we find that all galaxies would have been on top of one another something over 10 billion years ago. Since you can't get any closer together than that, this time must mark the beginning of the universal expansion, which is all we really mean by the Big Bang.

The expansion that began with the Big Bang has continued ever since, but the expansion rate has not always been the same in all places at all times. In particular, the force of gravity, which attracts all objects to all other objects, has presumably slowed the overall expansion, and in some relatively small regions of the universe gravity has halted and even reversed the expansion altogether. Those regions where gravity has halted the local expansion are the galaxies themselves—and also some groups and clusters of galaxies. Gravity also drove the collapse of smaller clumps of gas and dust within galaxies, thereby forming stars and planets. Galaxies like our own Milky Way probably looked much as they do today when the universe was less than half its present age.

Stars live and die within galaxies, expelling much of the matter from which they were made at the ends of their lives. This expelled material is then recycled into new generations of stars and planets. In this sense, galaxies function as cosmic recycling plants, providing the astronomical equivalent of ecosystems for stars, planets, and clouds of interstellar gas.

The earliest generation of stars had no Earth-like planets, because the Big Bang produced only the two simplest elements: hydrogen and helium (plus trace amounts of a third element, lithium). The rest of the elements, from which we and our planet are made, were manufactured in stars. Stars shine for most of their lives with the energy released by the nuclear fusion of hydrogen into helium. (Nuclear fusion is the process of making heavier elements from light ones by joining together nuclei of light elements.) But near the ends of their lives, advanced fusion reactions in the more massive stars produce all the remaining elements, including primary ingredients of life such as carbon, oxygen, nitrogen, and iron. That's how these elements came to be present in the universe, and it is from this "star stuff" (to quote the late Carl Sagan) that Earth and its life were made. The painting on the inside back cover summarizes this scientific story of our cosmic origins.

The ease with which we can state our cosmic address and origins hides some amazing ideas of scale. Interstellar distances are so vast that ordinary units of measure like miles or kilometers are almost useless. Instead, we usually measure cosmic distances in units of *light-years,* with one light-year being the distance that light can travel in a year. (If you prefer converting to ordinary units, a light-year is roughly 10 trillion kilometers, or 6 trillion miles.) Keep in mind that light travels extremely fast by Earth standards: it could circle the globe nearly eight times in just one second and takes only about six hours to journey from Earth to Pluto, the outermost planet of our solar system. But even at this incredible speed, light takes more than four years to travel the distance to the nearest star system, Alpha Centauri, which is why we say that Alpha Centauri is a little over four light-years away.

The fact that light takes significant amounts of time to traverse the vast expanses of space causes space and time to be inexorably intertwined. For example, Color Plate 1 shows a photograph of the Great Galaxy in Andromeda, also known as M31, which lies about 2.5 million light-years away. The photograph therefore shows how M31 looked about 2.5 million years ago, when early ancestors of modern humans were walking the Earth. Moreover, the diameter of M31 is about a hundred thousand light-years, so light from the far side of the galaxy required a hundred thousand years more to reach us than light from the near side. Thus, the photograph of M31 spans a hundred thousand years of time.

Ultimately, the finite age of the universe and the speed of light combine to limit the portion of the universe that we can see. For example, if the universe is 12 billion years old, then light from galaxies more than 12 billion light-years away would not yet have had time to reach us. We would therefore say that the *observable universe* extends 12 billion light-years in all directions from Earth. (This explanation is somewhat oversimplified, but the details are not important for our purposes.) The entire universe, meaning the whole of creation, is greater in size than the observable universe; in fact, as we'll discuss later, the entire universe might even be infinite in size.

The scale of time is no less incredible. The late Cornell University astronomer Carl Sagan pioneered the use of a wonderful device for comprehending astronomical time, which he called the cosmic calendar (below). We imagine the entire history of the universe compressed into a single year: the Big Bang occurs at the first instant of January 1, and the present day is the stroke of midnight on December 31. On this scale, the Milky Way galaxy probably formed sometime in February, but it was not until mid-August that our solar system was born. Life on Earth took hold by mid-September, but the great burst in diversity of life known as the Cambrian explosion did not occur until mid-December. The earliest dinosaurs walked the Earth on Christ-

JANUARY	FEBRUARY	MARCH	APRIL
Jan. 1	Milky Way		
Big Bang	forms		
MAY	JUNE	JULY	AUGUST
			Aug.13
			Earth forms
SEPTEMBER	OCTOBER	NOVEMBER	DECEMBER
			December...

Jan. 1: The Big Bang
February: The Milky Way forms
Aug. 13: The Earth forms

DECEMBER

S	M	T	W	T	F	S
1	2	3	4	5	6	7
8	9	10	11	12	13	14
15	16	17	18	19	20	21
22	23	24	25	26	27	28
29	30	31				

Dec. 13: Invertebrate life arose
Dec. 25: Dinosaurs walked the Earth
Dec. 30: Extinction of dinosaurs

December 31...

mas. Early in the morning of December 30, an asteroid or comet crashed to Earth, precipitating a global catastrophe that wiped out the dinosaurs and many other species of the time.

With the dinosaurs gone, small furry mammals inherited the Earth. Some 60 million years later, or around 9 PM on December 31 of the cosmic calendar, the earliest hominids (human ancestors) walked upright. Most of the major events of human history have taken place within the final seconds of the final minute of the final day on the cosmic calendar. With *now* being the stroke of midnight on December 31, agriculture arose only about thirty seconds ago. The Egyptians built the pyramids about thirteen seconds ago. It was only about one second ago on the cosmic calendar that Kepler and Galileo provided concrete proof that Earth is a planet orbiting the Sun. The oldest living humans were born less than three-tenths of one second ago. On the scale of cosmic time, the human species is the youngest of infants, and a human lifetime is a mere blink of an eye.

So here we are today, babes staring into the infinite unknown. Put yourself back on the shore of the mountain lake. Listen to the sounds of the night, and watch the reflections of the stars slowly circling overhead. Let your mind wander freely once again, contemplating the deepest questions of human existence. But this time consider the great foundation of knowledge that we have inherited through the efforts of thousands of our ancestors over thousands of years. This is their gift to you, a gift that enables you to join the human adventure of astronomical discovery. Now it is time to look ahead and see what mysteries have been laid at the doorstep for our generation to solve.

This version of the cosmic calendar assumes that the universe is 12 billion years old, so that each month represents about 1 billion years and each second represents about 380 years.

DECEMBER 31

1:00 PM
2:00 PM
3:00 PM
4:00 PM
5:00 PM
6:00 PM
7:00 PM
8:00 PM
9:00 PM — 9:00 PM Earliest human ancestors
10:00 PM
11:00 PM — 11:58 PM Modern humans evolved
12:00 PM

9:00 PM: Earliest human ancestors
11:58 PM: Modern humans evolved

60 seconds to midnight...

11:59:30 PM: Agriculture arose
11:59:47 PM: Pyramids built
11:59:59 PM: Kepler and Galileo proved Earth a planet

*Reality provides us with
facts so romantic that imagination itself
could add nothing to them.*

JULES VERNE (1828–1905)

Is There Life Elsewhere in Our Solar System?

Science fiction writers once wrote epic stories of civilizations on Mars. Visions of a Martian civilization have long since faded, but today scientists are debating whether a meteorite holds fossil evidence of microbial life on Mars. In this mystery, we will discuss the prospects for finding life elsewhere in our solar system, with particular emphasis on the cases of Mars, Europa (a moon of Jupiter that is believed to have a subsurface ocean), and Titan (a moon of Saturn that is rich in organic compounds).

A century ago, belief that a civilization lived on Mars was so widespread that the term "Martian" became essentially synonymous with "alien." Although occasional speculation about life on Mars goes far back into history, the craze began in 1877, when Italian astronomer Giovanni Schiaparelli reported seeing linear features across the surface of Mars through his telescope. He named these features *canali,* the Italian word for "channels." English accounts mistranslated the term as "canals," and coming amidst the excitement surrounding the recent opening of the Suez Canal (in 1869), Schiaparelli's discovery soon inspired visions of artificial waterways built by a Martian civilization.

Schiaparelli himself remained skeptical of such claims, and it's not clear whether he even thought the *canali* contained water. He may simply have been following a long-standing tradition adopted for the Moon, where visible features were referred to as bodies of water. But his work caught the imagination of a young Harvard graduate named Percival Lowell.

Lowell came from a wealthy and distinguished New England family. His brother Abbott became president of Harvard and his sister Amy was the well-known poet. After a few years as a businessman and as a traveler in the Far East, Percival Lowell turned to astronomy. Impassioned by his belief in the Martian canals and enabled by his wealth, Lowell commissioned an observatory (the Lowell Observatory) in Flagstaff, Arizona. Upon its completion in 1894, he set about making regular observations of Mars. In 1895, he published detailed maps of the canals, as well as a book expounding not only his belief that they were constructed by a Martian civilization but also his conclusions about the nature of that civilization. Because Lowell had correctly deduced that Mars was generally arid and had icy polar caps, he imagined that the canals were built to carry water from the poles to thirsty cities nearer the equator. (He was incorrect in assuming the polar caps to be water ice; they are largely carbon dioxide ice.) From there it was a short step to imagine the Martians as an old civilization on a dying planet, and the global network of canals convinced Lowell that they were citizens of a single, global nation. Such ideas provided the "scientific" basis for H. G. Wells's *The War of the Worlds,* published in 1898.

The canal myth persisted for decades, despite the fact that Lowell's canals never showed up clearly in photographs and other astronomers did not see them. Belief in Martians remained widespread enough to create a famous panic during Orson Welles's 1938 radio broadcast of *The War of the Worlds,* when many people thought an invasion was actually under way. The debate about Martian canals and cities was not entirely put to rest until NASA's first two spacecraft to Mars—the Mariner 4 flyby in 1965 and the Mariner 9 orbiter in 1971—sent back images of a barren, cratered surface.

Today we can be quite certain that there is no great Martian civilization, and probably never has been. But the possibility of microbial life on Mars remains a hot topic in science. Moreover, several other worlds in our solar system now seem like possible homes to indigenous life. Thus we are led to Mystery 10 on our list: Is there life elsewhere in our solar system?

(Above) This photograph shows a working model of the Viking landers, identical to those that landed on Mars in 1976. It is on display at the National Air and Space Museum in Washington, D.C. (Left) This pair of before and after photographs, transmitted back to Earth by the Viking 2 lander, shows where the robotic arm pushed away a small rock on the Martian surface.

The first step in approaching this mystery is deciding just what it is that we're looking for. Even on Earth, biologists are not quite sure how to define life. Plants and animals and bacteria are all clearly living, but what about viruses, which cannot even reproduce unless they invade some other organism? If we tried to imagine every conceivable form that life might take, the list might well be endless. We cannot address such infinite possibilities all at once, so the best way to approach this mystery is by searching for "obvious" life—life that bears enough resemblance to life on Earth that we will know it when we see it. This search began when Viking 1 landed on Mars on July 20, 1976.

Mars is an appealing place to look for life, because it is the most Earth-like planet in the solar system, at least on the surface. Nevertheless, today's Mars would not be a very comfortable place to visit. The atmosphere is far thinner than that on the top of Mount Everest, and no human could survive more than a few minutes without a pressurized spacesuit. The temperature is usually well below freezing, and there is no liquid water anywhere. And because the atmosphere contains only trace amounts of oxygen and hence essentially no ozone (which is a form of oxygen), the Sun's deadly ultraviolet rays pass unhindered to the surface. Any Earth-like organism that left itself exposed would be quickly killed by this ultraviolet radiation.

But photographs of Mars taken by orbiting spacecraft reveal unmistakable evidence of a very different past—dried-up riverbeds, vast floodplains, and perhaps even a huge ocean. Clearly, water once flowed on the Martian surface, and that could have happened only if Mars once had a much thicker and warmer atmosphere. Although other geological evidence suggests that any oceans dried up at least 3 billion years ago (but photographs from Mars Global Surveyor suggest that some water flows may be much more recent), we have no reason to doubt that conditions on the young Mars were nearly as hospitable to life as conditions on the young Earth. Because life on Earth was already flourishing more than 3.5 billion years ago, it's easy to imagine that the same thing occurred on Mars. If so, Martian life may have found ways to survive—perhaps underground or in the icy polar caps—even as Mars itself became a cold, dry desert.

Viking 1 and its twin, Viking 2, were each equipped to search for Martian life in a fairly simple way. (Each was also accompanied on its journey to Mars by a Viking orbiter, which studied Mars from above.) Like all planetary missions to date, the Viking spacecraft made only one-way journeys. Observations and experiments designed to search for life had to be planned before the spacecraft were launched and were carried out robotically by the spacecraft on Mars. The spacecraft then transmitted pictures and other scientific data from Mars back to Earth. The Viking landers could not move about the surface at all, but each lander had a robotic arm that it used to push small rocks aside and scoop up samples of Martian soil. A few simple, automated experiments then analyzed the soil, searching for organic compounds and signs of biological activity, such as respiration. Although the experiments did show some surprising results, planetary scientists soon explained these as chemical rather than biological reactions.

The negative results from Viking were surely disappointing to many scientists, but the experiments did not rule out life on Mars. The Viking missions had, after all, sampled soil at only two relatively bland locations on Mars, which had been chosen because they made relatively easy landing sites, and the search was carried out by only a few simple experiments. But budgetary and political considerations, the grounding of the space program for three years after the tragic Challenger accident in 1986, and the failure of two Russian missions (Phobos 1 and 2) and one American mission (Mars Observer) to Mars all conspired to stop Martian exploration for some twenty years. The new era began with the landing of

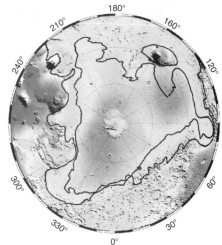

(Top) This Viking orbiter photo shows dried-up ancient riverbeds on the Martian surface. (Bottom) This projection shows the Northern Hemisphere of Mars, with the North Pole at the center. Darker areas represent higher elevations. The black curves represent possible shorelines of an ancient Martian ocean.

Mars Pathfinder and its little rover, Sojourner, on July 4, 1997 (see Color Plate 2).

Pathfinder, a simpler and considerably less expensive spacecraft than Viking, landed on Mars in a rather innovative way. A parachute slowed its descent through the Martian atmosphere, but it still hit the surface at crash-landing speed. To protect itself, Pathfinder deployed airbags on the way down, which completely encased the spacecraft components. These airbags allowed it to bounce along the surface until it finally came to rest. Over the next several months, Pathfinder surveyed the landing site with cameras and other on-board instruments. Sojourner, which had been stowed on Pathfinder for the journey from Earth, slowly roamed the neighborhood. (Sojourner was named for Sojourner Truth, an African American heroine of the Civil War era who traveled the nation advocating equal rights for women and blacks.) Although Sojourner could travel only a few meters from Pathfinder, this was enough to check the chemical composition of many nearby rocks. The results confirmed what scientists had suspected: the landing site, in the Ares Vallis region, is a vast, ancient floodplain. Rocks of many types lie jumbled together as they were deposited by the flood, and the departing waters left rocks stacked against each other in the same manner that floods do on Earth.

Pathfinder and Sojourner were designed more to test efficient and cheap new technologies than to search for life, but they marked the beginning of an ambitious invasion of Mars by spacecraft from Earth—and perhaps someday soon by human beings. As Earth and Mars orbit the Sun, the two planets are sometimes reasonably close (if 30 million miles can be called close) and other times much farther apart. When they are close, which happens roughly every two years, a journey to Mars takes about eight months with current technology.

NASA hopes to send two or more space probes to Mars at every one of these biennial opportunities over the next decade; however, the failure of both 1999 missions (the Mars Polar Lander and the Mars Climate Orbiter) may slow the parade somewhat while scientists and engineers figure out how to prevent similar losses in the future. Other nations, notably Japan and the member nations of the European Space Agency, also have plans for robotic missions to Mars. (The first Japanese mission to Mars, called Nozomi, was launched in 1998 but will not arrive until at least 2003, due to an engine misfiring shortly after launch.) The more sophisticated probes may carry rovers capable of journeying for many miles across the Martian surface in search of life.

Naturally, the search for life would be easier if we could bring a sample of Martian soil back to Earth, where we could subject it to analysis more detailed than is possible with automated probes that remain on Mars. NASA hopes to launch a sample-return mission to Mars as early as 2005, in which case we could have newly collected Martian rocks and soil in our hands as early as 2007 or 2008.

Surprisingly, that would not be our first sample of Mars, as scientists believe that a few meteorites may have a Martian origin. Almost all meteorites share common characteristics which tell us that they are either pieces of rock left over from the formation of our solar system or pieces of shattered asteroids. (Indeed, radioactive dating of the most ancient meteorites is what tells us that our solar system formed 4.6 billion years ago.) But a few meteorites have unusual compositions that suggest a different origin. These rare meteorites seem to be chunks of rock from the surface of the Moon and Mars. Based on what we know about the chemical composition of Mars from the Viking landers, fifteen of these meteorites (as of this writing) have compositions that are near perfect matches for what we expect in rocks from Mars.

In 1996, a team of NASA scientists claimed that one of these Martian meteorites, known as Allan Hills 84001, or ALH 84001 for short, contained evidence of past life on Mars. Meteorite ALH 84001 had been scooped from the Antarctic ice by a team of meteorite hunters in 1984. Careful study of the meteorite reveals its story. Radioactive dating shows that the underlying rock formed from volcanic lava, presumably on the surface of Mars, about 4.5 billion years ago. Sometime later—probably about 3.6 billion years ago—the rock was heated and deformed, so that a liquid was able to flow through it and deposit small rounded globules of material. Much more recently, the rock was heated again, this time by the impact of an asteroid that blasted it off

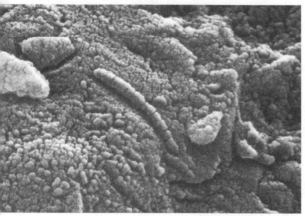

(Top) Meteorite ALH 84001 weighs a little less than 5 pounds and measures about 6 inches long by 4 inches by 3 inches. It is apparently a rock from Mars. (Bottom) Some scientists believe that the rod-shaped structures seen in this microscopic view of a thin section from ALH 84001 are fossils of microbial Martian life, but most others are skeptical.

the Martian surface. In space, the meteorite was exposed to cosmic rays—high-energy particles that leave telltale chemical signatures on anything unprotected by an atmosphere. Careful analysis of ALH 84001's cosmic ray exposure shows that it wandered through the solar system for at least 16 million years. Then, about thirteen thousand years ago, it came crashing down to Earth, landing in Antarctica.

The globules found in ALH 84001 contain chemical evidence suggestive of life, including layered carbonate minerals and complex molecules (polycyclic aromatic hydrocarbons)—both of which are generally associated with life when found in Earth rocks. Even more intriguing, under high magnification the globules reveal eerily lifelike, rod-shaped structures. However, while some scientists suggest that these structures might be fossils of microscopic Martian life, others argue that all the so-called evidence for Martian life in ALH 84001 could have been produced by nonbiological processes.

This extraordinary debate about whether we already have discovered evidence of life beyond Earth will undoubtedly continue as scientists subject ALH 84001 and other Martian meteorites to further scrutiny. Unfortunately, unless there is a surprising and major breakthrough, it is unlikely that the meteorites by themselves can provide definitive evidence of life. For that, we will have to rely on the planned armada of spacecraft bound for Mars.

Not long ago, most scientists would have said that Mars was not only the best place to look for life in our solar system but the *only* place. Our Moon and the planet Mercury are both cratered worlds without atmospheres, and clearly could not harbor life as we know it. Venus is a near twin to Earth in size, but its thick carbon dioxide atmosphere creates a strong greenhouse effect that bakes its entire surface to 450°C (900°F)—far hotter than a pizza oven. In addition, the weighty atmosphere bears down on the surface with a pressure equivalent to that nearly a kilometer beneath the ocean surface on Earth, and the Venusian clouds contain sulfuric acid and other corrosive chemicals. Such conditions make it difficult to imagine life on Venus. With Mercury, Venus, and the Moon ruled out, and Mars already considered, we are left with the worlds of the outer solar system. These worlds and their moons are magnificent, but they are very far from the Sun. Whereas Mars orbits the Sun only about half again as far away as Earth, Jupiter orbits at five times the Earth's distance, and the other outer planets (Saturn, Uranus, Neptune, and Pluto) are much farther still. Sunlight is too weak to provide the energy needed for life on these distant worlds. But who needs sunlight?

Until fairly recently, most scientists would have answered, "All life forms do." But the past few decades have seen extraordinary discoveries about the abodes of life on Earth. Biologists have found microorganisms living deep inside rocks in the frozen deserts of Antarctica and inside rocks buried nearly a mile underground. Perhaps most important to the search for life beyond Earth, scientists have found teeming life near deep undersea volcanic vents. This life does not in any way depend on sunlight, instead receiving all its chemical energy from the heat of the Earth itself. Moreover, genetic analysis of life-forms around the volcanic vents suggests that they may be more closely related to the common ancestor of all life on Earth than any other living organisms. Many biologists therefore believe that life may have first arisen around volcanic vents, with sunlight-dependent life coming only later. If so, then the water around undersea volcanoes might be

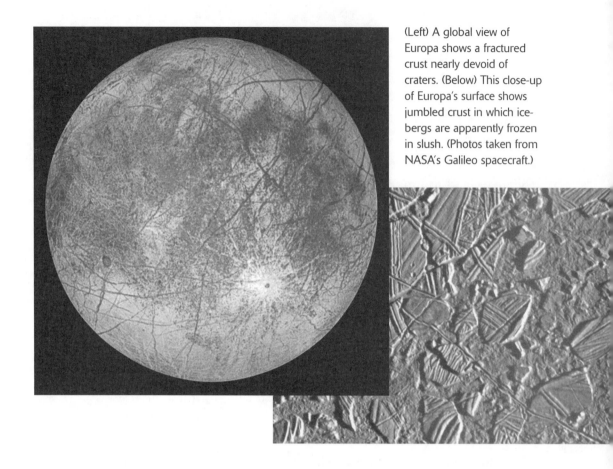

(Left) A global view of Europa shows a fractured crust nearly devoid of craters. (Below) This close-up of Europa's surface shows jumbled crust in which icebergs are apparently frozen in slush. (Photos taken from NASA's Galileo spacecraft.)

the best place to find life in our solar system—and we know of at least one place beyond Earth where such volcanoes may be found.

Europa is one of four large moons that orbit Jupiter, along with a dozen smaller moons. Europa was first seen through a telescope by Galileo, but we had no close-up pictures of it until the Voyager 1 and 2 flybys in 1979. Photographs of Europa show a surface with very few impact craters. In a solar system filled with rocky debris, a nearly crater-free surface can mean only one thing: some process is continually erasing the evidence of impacts. Because Europa's surface is made almost entirely of water ice, it seems likely that the craters have been erased by flowing water. Europa is therefore thought to have a huge sub-surface ocean beneath its icy crust. Recent, detailed photographs taken by the Galileo spacecraft—which has been orbiting Jupiter since

1995—support this basic scenario. Moreover, the structure of the surface ice suggests at least occasional heating from below, presumably by volcanoes sprouting from the floor of Europa's ocean (see Color Plate 3).

With a deep ocean and a source of energy in its undersea volcanoes, Europa may be even more likely to harbor life than Mars. Indeed, if biologists are correct in guessing that life on Earth first evolved near undersea volcanic vents, an absence of life on Europa may be more difficult to explain than its presence.

Not surprisingly, Europa is now a prime target for scientific exploration. NASA is currently developing plans for a Europa orbiter that could be launched as early as 2003 to reach Europa in 2006 or 2007. This mission would carry a radar instrument designed to establish whether or not Europa really has a subsurface ocean, and perhaps how much liquid water lies beneath the icy crust. If the existence of an ocean is verified, a follow-on mission would send a spacecraft to land on Europa, where it would use battery-generated heat to melt its way through the crust and into the ocean. There's no telling what such a mission might find, from microbial life to creatures larger than whales.

Europa provides the clearest case to date for a subsurface ocean, but Ganymede and Callisto, two other moons of Jupiter, may also have such oceans. These worlds, too, have icy surfaces, but their larger numbers of craters show that water flows occur less commonly, if at all. Nevertheless, it's certainly worth looking for life on both of these moons as well. Imagine the irony if we found life on all three. There might be more inhabited worlds around the far-flung planet Jupiter than in the rest of the solar system combined.

Hopes for finding life dim beyond Jupiter, but a few scientists hold out the possibility of life on Saturn's moon Titan. Although we usually think of moons as being airless, Titan has a thick atmosphere—like Earth's, it is made mostly of nitrogen. The surface pressure is only slightly greater than that on Earth, which means that it would be fairly comfortable if not for the lack of oxygen and the bone-chilling temperatures. Titan is far too cold for liquid water today, but the heat of impacts early in its history might have created slushy ponds in which life potentially could have arisen. If life did arise, perhaps some organisms survive to the present day in geological hot spots or somehow adapted to survive despite Titan's icy temperatures. We do not yet know what Titan's surface looks like, because the thick atmosphere hides it from view. However, the combination of atmospheric composition and surface temperature, both of which have been measured, leads many scientists to believe that Titan may be partially or fully covered by a deep ocean of liquid ethane.

Answers about Titan may be coming fairly soon, because a spacecraft called Cassini is scheduled to arrive there in 2004. Cassini was launched in 1997 and carries a radar instrument that should allow us to map Titan's surface in detail. It also carries a probe called Huygens (named for the seventeenth-century astronomer who discovered Titan), which it will drop to the surface of Titan. Just in case, the probe is designed to float in liquid ethane.

We have now identified five places in our solar system with a reasonable chance of harboring Earth-like life: Mars, Europa, Ganymede, Callisto, and Titan. If we allow for less likely locations, the list expands even more. The four giant planets of our solar system—Jupiter, Saturn, Uranus, and Neptune—have no solid surfaces upon which to look for life, but their atmospheres are rich in the chemical building blocks of life, and at some depths the temperature would be quite comfortable for life. Although it is difficult to imagine how life might arise in the air in the first place, once it got started life might survive just fine—as long as it found some way to stay at a comfortable depth despite the turbulent atmospheric currents.

A more speculative idea suggests that life might exist on comets, which we know to contain large amounts of organic material. A few scientists have even suggested that life on Earth may have been brought here by a comet. If so, comets may have seeded life on many other worlds. Moreover, the presence on Earth of meteorites from Mars tells us that planets occasionally exchange rocks dislodged by major impacts. The harsh conditions under which some life on Earth exists suggest that living organisms might survive such impacts and the long journey from one planet to another. In that case, life may have migrated among

The inset (right) is a photograph of part of Titan from Voyager 2. Note that it is completely enshrouded by its thick atmosphere. The artist's conception (above) shows what the surface might look like as the Huygens probe descends in 2004. Saturn and the Cassini orbiter are faintly visible through the atmosphere. (Artistic license has been taken to make the orbiter look big enough to be seen from the moon's surface.)

the planets just as various species spread among islands on floating debris. Life from Earth may have made its way to Mars—or vice versa. If we ever discover life on another world, we should be able to tell whether it shares a common origin with life on Earth by examining its set of genetic instructions.

The search for life may well be a never-ending quest. If we don't find it, we can always believe that we simply haven't yet looked hard enough or in the right places. But whether or not we find life, the exploration of Mars, Europa, and Titan planned for the next decade should provide deep insights into the conditions under which life can survive. And if this coming exploration does find life on other worlds, our Mystery 10 will be solved before the third millennium is barely under way.

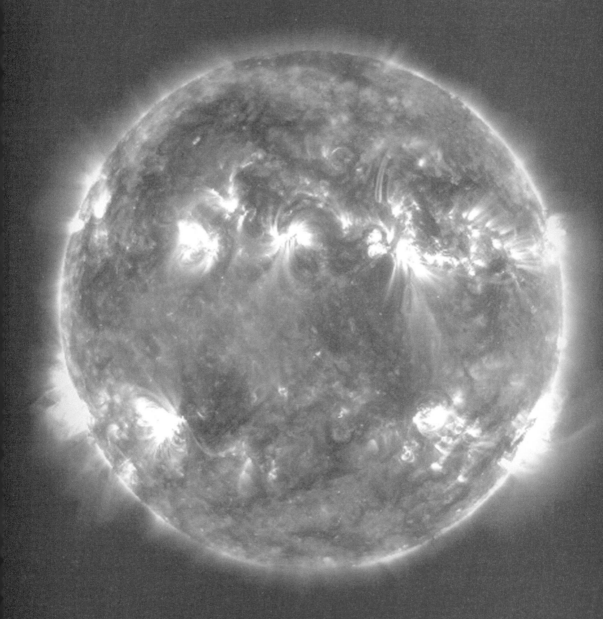

I say Live, Live, because of the Sun,

The dream, the excitable gift.

ANNE SEXTON (1928–1974)

Where Are the Sun's Neutrinos?

The Sun is the source of nearly all light and heat on Earth. It generates this energy through nuclear fusion in its core. Astronomers believe they understand the fusion process very well, and this understanding lies at the heart of all other astronomical understanding of stars. But there is one nagging problem: thermonuclear fusion in stars is known to generate subatomic particles called *neutrinos,* and experiments over the past four decades have indeed detected neutrinos coming from the Sun—but in significantly fewer numbers than theory predicts. Most astronomers believe the solution to this problem lies not with any misunderstanding of the Sun or of fusion in stars, but instead with our understanding of neutrinos. The resolution to the solar neutrino problem may thus turn out to have deep implications for our understanding of the universe as a whole.

Here's a mystery that spanned several millennia: Why does the Sun shine? Most ancient thinkers speculated that the Sun was some type of fire, perhaps a lump of burning coal or wood. The Greek philosopher Anaxagoras (c. 500–428 BC), having heard of a meteorite fall on the shore of the Aegean Sea, concluded that the Sun must be a very hot, glowing rock. He guessed its size to be about that of the Greek peninsula of Peloponnesus (comparable in size to Massachusetts). No significantly better ideas came up for the next two thousand years, although the later Greek astronomer Aristarchus at least realized that the Sun is larger than the Earth. But after Copernicus, Kepler, Galileo, and Newton uncovered the mysterious mechanics of the heavens, scientists were gradually able to determine the Sun's true size and distance with ever increasing accuracy.

By the mid-1800s, the size and distance of the Sun were reasonably well known, but the question of how the Sun shines remained as mysterious as ever. Simple calculations showed that even with its enormous size—the Sun's mass is greater than three hundred thousand Earths—no known energy source could explain its power. If scientists assumed that the Sun generated energy chemically by burning like coal or wood, they found that it was not physically large enough to account for the rate at which sunlight was leaving its surface. An alternative proposal held that the Sun had been created extremely hot and now was a cooling ember, but this idea failed because just a few hundred years earlier the Sun would have been too hot for human life on Earth.

For a while, most astronomers were enamored of an idea proposed by the famous British physicist William Thomson (later known as Lord Kelvin) and the German physicist Hermann von Helmholtz in the 1850s. They suggested that the Sun generates energy by gradually shrinking in size. Just as a dropped rock converts energy that it has by virtue of its height (called *gravitational potential energy*) to energy of motion as it falls to the ground, the process of shrinking should liberate energy in a star. Detailed calculations showed that the Sun could have generated energy by this mechanism for as long as 20 million years, which greatly pleased most astronomers. Unfortunately, the geologists and biologists of the time were far less impressed, as their work showed the Earth and, indeed, life to be at least hundreds of millions of years old. Surely the Earth could not be older than the Sun!

The mystery of the Sun's energy was finally solved in the first half of the twentieth century. In 1905, Einstein published his famous equation $E = mc^2$, which shows that mass is just a special form of energy. Although no one yet knew exactly how mass might be converted to other forms of energy, it was easy to see that even a small fraction of the Sun's mass contained enough energy to account for the Sun's shining for billions of years. In 1920, relying on recent measurements of the mass of hydrogen and helium nuclei, the British astronomer Arthur Eddington suggested that the Sun generates energy by fusing hydrogen into helium in its core. (In the same paper, he presciently wrote that we might someday harness this power ourselves "for the well-being

1

M31, the Great Galaxy in Andromeda. It is about 2.5 million light-years away, so this photograph captures light that traveled through space for 2.5 million years to reach us. Because the galaxy is 100,000 light-years in diameter, the photograph also captures 100,000 years of time in M31. We see the galaxy's near side as it looked 100,000 years later than the time at which we see the far side.

2

Panoramic view of a Martian floodplain from the Pathfinder lander, now called the Carl Sagan Memorial Station. Notice the rover Sojourner (about the size of a microwave oven) examining a nearby rock.

3 (next page)

This painting shows what the surface of Europa may look like in a region where the crust has been disrupted by an undersea volcano. (Painting by Joe Bergeron.)

of the human race—or for its suicide.") Less than twenty years later, theoretical physicist Hans Bethe worked out the details of fusion reactions in the Sun. As a result, we can now explain the Sun's energy very simply: every second, the Sun fuses 600 million tons of hydrogen into 596 million tons of helium, with the "missing" 4 million tons turning into the pure energy that makes the Sun shine.

There's only one small problem with this scenario: we cannot see into the core of the Sun, and thus it is based almost entirely on theoretical considerations and indirect evidence. But there is one way to probe the Sun's core directly, and it leads us to our Mystery 9, commonly known to scientists as the "solar neutrino problem."

Neutrinos are tiny subatomic particles that get their name, which means "little neutral ones," from the fact that they are electrically neutral and extremely lightweight. Until recently, they were assumed to be massless, and even today their precise mass is not known; however, the mass of a neutrino is certainly far less than the mass of an electron. (Electrons weigh only about one two-thousandth as much as the protons and neutrons that account for nearly all the mass of ordinary matter.) Neutrinos are what physicists call "weakly interacting" particles, because they rarely interact with each other or any other type of matter. This property gives them a ghostlike quality that makes them extremely difficult to detect.

According to theory, neutrinos are produced by many types of nuclear reactions—for example, by reactions in nuclear power plants, by thermonuclear fusion in stars, by the stellar explosions called supernovae, and by the reactions that created matter in the Big Bang. The neutrinos from the Big Bang ought to be very common in the universe, but these neutrinos have such low energies compared to neutrinos from other sources that they are completely undetectable by our present technology. Neutrinos from a supernova have been detected once, from the 1987 event known as Supernova 1987A.

Because the Sun generates its energy by nuclear fusion, it must produce neutrinos in prodigious numbers. As you read this sentence, about 1,000 trillion solar neutrinos will zip through your body. Fortunately, this flood of neutrinos is completely harmless. Neutrinos are so elusive that if you wanted to catch one, on average you would need a slab of lead more than a trillion miles thick. The neutrinos produced in the Sun's core fly outward through the Sun's dense interior more easily than you can sweep your hand through air. Traveling at close to the speed of light, those neutrinos that are headed in our direction reach Earth in a mere eight minutes.

The trick is to catch the neutrinos once they arrive. Nearly all of them will pass right through the Earth—just as they passed through the much larger and denser solar interior—continuing uninterrupted on their journey through space. But neutrinos do occasionally bang into other bits of matter, and since the 1960s astronomers have set up several experiments that catch a few solar neutrinos here and there. And therein lies the problem: the experiments catch fewer neutrinos than they should, if we are correct about the number coming from the Sun.

The first experiment designed to detect solar neutrinos began in 1967 and essentially relied on a giant vat of 100,000 gallons of dry-cleaning fluid. Dry-cleaning fluid contains chlorine, and on very rare occasions a chlorine nucleus will capture a neutrino and change into a nucleus of radioactive argon. By looking for the decay of radioactive argon nuclei, experimenters could count the number of neutrinos that had been captured. Other cosmic particles can cause the same reaction, so to prevent confusion the vat of cleaning fluid was placed nearly a mile underground, in the Homestake gold mine in South Dakota. Only the elusive neutrinos can easily penetrate a mile of solid rock.

Remarkably, from the many trillions of solar neutrinos that pass through the tank of cleaning fluid each second, experimenters expected to capture an average of just one neutrino per day. This predicted capture rate was based on measured properties of chlorine nuclei and models of nuclear fusion in the Sun. However, over a period of nearly three decades, neutrinos were captured an average of just about once every three days. That is, the Homestake experiment detected only about a

This tank of dry-cleaning fluid (visible beneath the catwalk) lay nearly a mile underground in the Homestake gold mine in South Dakota. It was the first dedicated solar neutrino detector.

third of the predicted number of neutrinos. This stunning result can mean only one of two things:

1. We are wrong about the number of neutrinos produced in the Sun, which means that something is wrong with our theory about how the Sun produces energy.
2. There's something strange about neutrinos themselves that we don't yet understand.

If science were a democracy, astronomers would vote immediately for option 2, thereby brushing aside any concerns about the supposedly answered mystery of how the Sun shines. But nature dictates the truth, and the job of science is to find it. For that, we need more data. As the scientists behind the Homestake experiment knew from the start, their cleaning-fluid detector was not very sensitive.

To understand what we mean by "sensitive" in this context, we need to delve a little more deeply into the mechanism of nuclear fusion in the Sun. Hydrogen fusion converts four hydrogen nuclei into a helium nucleus. A normal hydrogen nucleus is just a single proton, while a helium nucleus consists of two protons and two neutrons. Fusion releases energy because, as proved by careful measurements, a helium nucleus weighs slightly less than four hydrogen nuclei combined. (It was precisely such measurements that led Eddington to suggest hydrogen fusion as the Sun's source of power.) The "missing" mass in the helium becomes energy in accord with Einstein's equation $E = mc^2$, in which E stands for the energy released, m stands for the amount of mass converted to energy, and c stands for the speed of light. Thus hydrogen fusion essentially entails making four protons stick together while turning two of them into neutrons and releasing energy in the process:

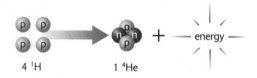

However, fusion does not occur in a single step. It proceeds through a series of steps, and neutrinos are by-products of some of them. Neutrinos produced by differing steps have differing amounts of energy. Some 99 percent of the time, the series of steps begins with two protons fusing together to make a deuterium nucleus, releasing a low-energy neutrino in the process. Higher-energy neutrinos are produced only by rare steps that occur in some of the remaining 1 percent of fusion reactions. Homestake's sensitivity problem arose because the chlorine in its detector could capture only these high-energy neutrinos, and thus told us nothing at all about 99 percent of the fusion reactions in the Sun.

Given the dramatic challenge to theory posed by the solar neutrino problem, it seemed a good idea to look for the more common but lower-energy neutrinos. After all, based on the Homestake experiment alone, it was at least possible that the problem was nothing more than a slight error in our understanding of a rare reaction pathway. Thus, as the Homestake results became clear, astronomers sought to build more sensitive detectors.

For scientists who understood neutrinos, the obvious choice was a tank containing the element gallium rather than chlorine. Gallium captures neutrinos in a manner similar to chlorine—except the product is radioactive germanium rather than radioactive argon—but it is much more sensitive to neutrinos and therefore able to capture the low-energy neutrinos from the main fusion reactions in the Sun. However, in the 1970s, scientists could not get their wish, because gallium was too rare and expensive. Fortunately for neutrino science, it soon became a favored material in the growing computer industry, which led to increased gallium mining. In the early 1990s, two gallium-based solar neutrino detectors came on line: SAGE (Soviet-American Gallium Experiment) in the Caucusus, and GALLEX (GALLium EXperiment) in the Gran Sasso mountains of Italy.

Like Homestake, the two gallium experiments found a lower number of solar neutrinos than expected—detecting between one-third and two-thirds of the predicted number. (The deficit is given as a range rather than a precise number because of experimental uncertainty.) Because this shortfall involves the solar neutrinos thought to be most common, it rules out the possibility of the deficit as a minor problem in our understanding of a rare fusion reaction in the Sun; in other words, the gallium results proved that a "solar neutrino problem" really does exist.

Meanwhile, a completely different type of neutrino detector was built in Japan. Located deep in a mine near Kamioka, this detector was originally built in the early 1980s to look for an effect unrelated to the solar neutrino problem: the decay of protons, as predicted by "grand unified theories" of physics. The original detector, called Kamiokande (Kamioka Nucleon Decay Experiment), consisted of a tank containing 4,500 tons of ultrapure water. In 1996, Kamiokande was replaced by an even larger detector, known as Super-Kamiokande or "Super-K," containing 50,000 tons of water (see Color Plate 4). More than 11,000 large phototubes line the tank of Super-K, searching for telltale flashes of light from neutrino (and other) interactions.

Super-K (and Kamiokande before it) can detect only high-energy solar neutrinos, which means it suffers from a sensitivity problem similar to that of the Homestake experiment. However, it offers one major advantage over both Homestake and the gallium experiments: it can determine precisely when a neutrino strikes and what direction it came from. In Homestake and the gallium experiments, scientists could count neutrino captures only after the fact, when they searched the tanks for the radioactive by-products left behind. In Super-K, the evidence is much more direct. When a solar neutrino strikes an electron in the water tank, it causes the electron to recoil at a speed faster than the speed of light in water (but slower than the speed of light in a vacuum, and therefore consistent with Einstein's theory of relativity). This generates a flash of light (called Čerenkov radiation) that is essentially the optical equivalent of a sonic boom. Since the phototubes detect this flash immediately, researchers know precisely when the neutrino hit the electron. The direction in which the electron moves through the water tells the researchers the direction from which the neutrino came.

Five hundred days' worth of data from the Super-K experiment allowed astronomers to construct this first-ever "neutrino image" of the Sun. Brighter colors correspond to larger numbers of neutrinos. The solar neutrinos all come from the Sun's core, but this image represents a much larger portion of the sky because the experiment does not give perfect directional information. Nevertheless, it offers clear proof that neutrinos are coming from the Sun.

The Super-K results further confirm the existence of the solar neutrino problem in two ways. First, the directional data prove that most of the detected neutrinos really do come from the Sun, rather than from somewhere else in space. Second, Super-K finds roughly half the number of neutrinos expected, a shortfall consistent with the other experiments.

With the solar neutrino problem definitively established as a real and major mystery, we're ready to consider its two possible solutions. The first, that we're wrong about the number of neutrinos that should be coming from the Sun, would be hard for scientists to swallow. Assuming that our basic understanding of nuclear fusion is correct—and the successful detonations of H-bombs in the 1950s and 1960s make it hard to imagine that it is not—the only way we can explain a shortfall of neutrinos is by a shortfall in the number of nuclear reactions taking place in the Sun.

The predicted number of fusion reactions comes from what astronomers call the *standard solar model,* in which basic laws of physics are used to calculate the temperature and pressure throughout the Sun's interior. If this model is wrong and the temperature is slightly lower than we think, the number of fusion reactions would be smaller and the solar neutrino problem would go away. However, during the 1990s astronomers began to probe the interior of the Sun in a new way: by observing "sunquakes," or vibrations of the Sun's surface similar to the vibrations produced on Earth's surface by earthquakes. The study of these sunquakes is called helioseismology (*helios* means "Sun"). Just as seismometers on Earth's surface provide geologists with information about Earth's interior, observations of the Sun's surface vibrations provide astronomers with information about the Sun's interior. These observations show that the Sun's interior structure nearly perfectly matches the predictions made by the standard solar model, essentially ruling out the possibility of a lower temperature in the solar core. Thus a "solar" solution to the solar neutrino problem would have to overturn almost everything we think we know about solar and fusion physics.

The preferable solution to the solar neutrino problem therefore lies with the neutrinos themselves, since it would keep intact our understanding of nuclear fusion in the Sun. In that case, we are correct about the number of neutrinos being produced by the Sun, and the experimental shortfall must come either because of errors in the experiments or because the "missing" neutrinos are somehow hiding. The fact that five different experiments using three types of detector (chlorine, gallium, and water) all give similar results makes it highly unlikely that the problem lies with the experiments. Thus most scientists now believe that the solution to the solar neutrino problem lies with neutrinos that can hide!

The idea of neutrinos hiding is not quite as bizarre as it might at first sound. While we have so far talked about neutrinos as though they were all the same, they in fact come in three different varieties, known as electron, muon, and tau. The theory of nuclear fusion tells us that all the neutrinos produced in the Sun are electron neutrinos, and the five experiments we've described can detect only this variety of neutrino from the Sun (though, as we'll discuss shortly, Super-K occasionally detects muon neutrinos that happen to have much higher energies than solar neutrinos). Thus any solar neutrinos that somehow transformed themselves into one of the other two kinds before they reached Earth would be effectively hidden.

The possibility of one sort of neutrino changing into another actually arose more than a decade before the discovery of the solar neutrino problem. It was first proposed in 1957 by physicist Bruno Pontecorvo (perhaps better known for having defected from Canada to the Soviet Union), based on considerations from quantum physics. According to quantum physics, subatomic particles have properties that we normally associate with both particles and waves. The wave nature of subatomic matter leads to many phenomena that seem strange in our everyday view of the world, such as quantum tunneling and the electron as a "cloud" surrounding the nucleus. Most relevant to neutrinos is the idea that waves can be superimposed on one another in a way that mixes their underlying identities. For example, when you play two keys on a piano at the same time, their sound waves mix so that your ears hear a sound different from that produced by either key individually. Similarly, it is conceivable that what we usually think of as an individual neutrino really consists of some mixture of two or more different neutrino types. If so, the identity of the neutrino at any given moment might oscillate among them. What is born in the Sun as an electron neutrino might therefore be some other type of neutrino when it strikes a detector on Earth.

Do neutrinos really oscillate among different identities? In mid-1998, scientists using the Super-K detector announced convincing evidence that this is indeed the case. Interestingly, the evidence came not from solar neutrinos, but from neutrinos produced in the Earth's atmosphere. The high-energy particles known as cosmic rays constantly bombard the Earth from distant reaches of space. (Their precise origin is unknown, but they probably are associated with supernovae.) When a cosmic ray hits an atom in the Earth's atmosphere, it starts a cascade of nuclear reactions, some of which produce neutrinos. Unlike solar neutrinos, these atmospheric neutrinos are of the muon type—at least initially. Moreover, they have very high energies compared to neutrinos from the Sun, which allows Super-K to detect them even though they are not of the electron type. Super-K measurements found that substantially more of these atmospheric muon neutrinos hit it from above than from below; that is, Super-K detects fewer muon neutrinos produced on the far side of the Earth than on its own side of the Earth. Since neutrinos pass through the Earth essentially unimpeded, the only explanation for the deficiency is that the longer distance traveled by neutrinos from the far side gives them time to transform themselves into a neutrino type that Super-K cannot detect.

In principle, now that we know neutrinos can oscillate among their types, it should be easy to determine whether these oscillations are the solution to the solar neutrino problem by building a detector that can detect neutrinos of all three known types at the energies of neutrinos from the Sun. If the missing solar neutrinos have been disguising themselves, this detector would catch them red-handed. Such a detector began operations in mid-1999, housed some 6,800 feet underground in a mine near Sudbury, Ontario. The Sudbury Neutrino Observatory, or SNO for short, is similar in design to Super-K; however, it uses heavy water rather than ordinary water as its detector. Heavy water is water in which the two hydrogen atoms are replaced by deuterium (making it D_2O instead of H_2O).

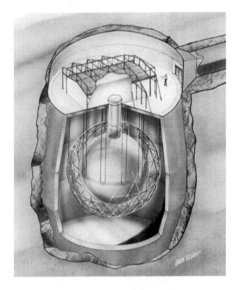

This diagram shows the design of the Sudbury Neutrino Observatory (SNO). The central sphere, made of acrylic, is 12 meters in diameter and holds 1,000 tons of heavy water. It is surrounded by 9,600 phototubes in a geodesic array. This entire setup is immersed in ordinary water (to help shield it from cosmic rays) in the large, barrel-shaped cavity excavated from the rock in the mine. SNO is located at the bottom of a mine shaft 6,800 feet underground.

Whereas an ordinary hydrogen nucleus consists only of a proton, a deuterium nucleus also contains a neutron and hence is heavier. Heavy water exists naturally, but separating it from ordinary water is difficult and expensive. SNO was made possible by the fact that Canada (unlike the United States) uses heavy water in its commercial nuclear reactors. Atomic Energy of Canada therefore has developed a stockpile of heavy water, and the company agreed to loan 1,000 tons of it—valued at over $200 million—to make SNO possible. Eventually the heavy water will be returned to its owner.

The deuterium nuclei in SNO's heavy water can capture all three known types of neutrino. Such captures break a deuterium nucleus apart, and special sensors in SNO can then detect the ejected neutron. SNO can also detect electron neutrinos separately in a manner similar to that used in Super-K. Thus SNO should be able to compare the number of electron neutrinos coming from the Sun to the *total* number of neutrinos coming from the Sun. With luck, the total number of solar neutrinos will match that predicted by the standard solar model, and the solar neutrino problem will be solved.

However, few scientists believe that the solar neutrino problem will be solved quite so simply and quickly. It is likely to be several years before SNO has enough data so that it can be carefully calibrated and the results can be properly interpreted. Moreover, careful analysis of the data from the five previous neutrino detectors has revealed some inconsistencies that SNO cannot by itself resolve—one reason why several other solar neutrino observatories are currently planned or under construction. (Another reason is the unremitting need to confirm important results in science.) Worse yet, one interpretation of the existing data implies the existence of undiscovered neutrino types that cannot interact with other matter *at all*. Because these so-called sterile neutrinos would be impossible to detect, we'd have no hope of ever counting any solar neutrinos that oscillate into such a form. In that case, neither SNO nor any other neutrino observatory would be capable of confirming that we are correct about what goes on inside the Sun.

Besides rendering the solar neutrino problem unsolvable, sterile neutrinos would give physicists the willies. After all, a fundamental tenet of all science is that we can use theories to make testable predictions. By being undetectable, sterile neutrinos would violate this tenet, making it impossible to know whether they really exist or are just a fantasy rigged up to save our theories. The betting at present is that sterile neutrinos won't be necessary to explain the experimental results, and that SNO plus the other new experiments will eventually put the solar neutrino problem to rest.

Even if we solve the solar neutrino problem without having to invoke sterile neutrinos, the fact of neutrino oscillations still has far-reaching implications. According to our present understanding of physics, neutrinos can transform themselves from one type to another only if they have a slightly greater mass when existing in one type than in the other. The Super-K finding of neutrino oscillations therefore proves that at least one type of neutrino does have mass, and it's possible that neutrinos of all three types have mass. The mass of any sort of neutrino is undoubtedly quite small, but any mass at all may be important for two reasons.

First, since the 1960s, physicists have built a theory about the fundamental nature of subatomic particles that has been so successful in explaining experimental results that it is known as the *standard model of particle physics* (not to be confused with the *standard solar model,* which applies to the structure of the Sun). However, in its current state, the standard model of particle physics assumes that neutrinos have no mass. Thus the discovery of neutrino mass has forced physicists to revisit the model, looking for a way to patch it up or replace it. The revision must still explain all the experimental results consistent with the current model while also allowing for neutrinos with mass.

Second and more important for astronomy, neutrinos are thought to be very common. The Big Bang theory predicts that neutrinos far outnumber ordinary particles such as protons, neutrons, and electrons. Thus, even with very tiny individual masses, the sheer number of neutrinos could make them a significant fraction of the total mass of the universe. In that case, as we will discuss in Mystery 2, neutrinos could play a major role in determining the fate of the universe. The search for a solution to the solar neutrino problem may be over within a decade or two, but the implications of its solution may well tell us how the universe will behave until the end of time.

The great mystery is not that we should
have been thrown here at random
between the profusion of matter and that
of the stars; it is that from our very
prison we should draw, from our
own selves, images powerful enough
to deny our nothingness.

ANDRÉ MALRAUX (1901–1976)

What Does the Universe Look Like?

We live on a planet that orbits an average star in the Milky Way galaxy. The Milky Way, in turn, is just one of billions of galaxies in our universe. But how are these galaxies arranged? Are they strewn randomly about, or are they small pieces of much larger structures? Many people are surprised to learn that we still have so much to learn about this question. In this mystery, we will discuss why knowing the large-scale structure of the universe is so important to our overall understanding of the cosmos and why it is difficult to determine this structure. We will also discuss how we've reached our current understanding and how much of the remaining mystery of large-scale structure may be resolved over the next decade.

Imagine for a moment that the time is a few thousand years ago and you live in a small village. The ground beneath you feels solid and un-moving, and your conception of up and down tells you that aside from hills and valleys the world must be flat. The sky appears to be a dome upon which the Sun makes its daily circuit, and which fills with stars at night. In short, you would have a clear and simple picture of your universe as a small, flat Earth covered by the dome of the sky.

One day, you embark on a trek of exploration, knowing that you may be leaving friends and family forever. Upon reaching a village on the coast, you join the crew of a small ship, voyaging to distant shores. Clearly, the world is much larger than you had imagined. Moreover, you begin to notice different stars as you travel north or south, and conversations with fellow travelers confirm that the "dome" of the sky must be more than a simple hemisphere.

It was evidence like this that convinced the Greek scientist Anaximander (c. 610–547 BC) that the stars must fill a great celestial sphere. He also realized that the changes in the sky seen by travelers meant that the Earth must be curved in a north-south direction, and he therefore imagined the Earth as a cylinder lying in the center of the celestial sphere. By about 500 BC, the additional evidence provided by Earth's curved shadow on the Moon during lunar eclipses convinced the famous mathematician Pythagoras that the Earth, like the heavens, must be a sphere. This Pythagorean idea of the universe as a celestial sphere surrounding a central, spherical Earth remained largely intact for the next two thousand years. Of course, this view of the universe was accessible only to the relatively few educated people. Most of the public remained illiterate and never learned of ideas beyond the seemingly obvious notion of a flat Earth under a dome-shaped sky.

The idea of a fairly small, Earth-centered universe came under serious attack when Nicolaus Copernicus published in 1543 his revolutionary idea about the Earth going around the Sun. This idea was not entirely new; as we discussed in the Introduction, a Sun-centered cosmos had been proposed by Aristarchus some eighteen hundred years earlier. Copernicus was aware of this ancient proposal and took it several steps farther by creating a detailed mathematical model to describe how the planets moved in their orbits around the Sun. (Recall that Nicholas of Cusa also suggested a Sun-centered cosmos about a century before Copernicus, but he did not offer a detailed model and it is unlikely that Copernicus ever heard of his work.) However, other ancient Greeks had pointed out what appeared to be a fatal flaw in the proposal by Aristarchus, and this was a major reason why the idea of the Earth as a planet had never caught on. The supposed flaw involved something called *stellar parallax*.

Extend your arm and hold up one finger. If you keep your finger still and alternately close your left and right eye, your finger will appear to jump back and forth against the background. This apparent shifting is called *parallax* and occurs simply because your two eyes view your finger from opposite sides of your nose. If you now imagine that

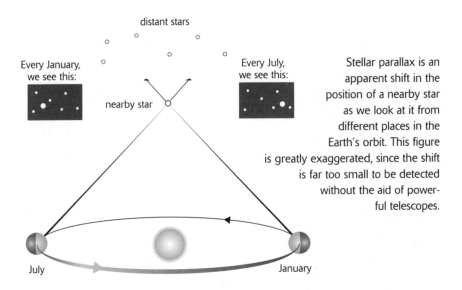

distant stars

Every January, we see this:

nearby star

Every July, we see this:

Stellar parallax is an apparent shift in the position of a nearby star as we look at it from different places in the Earth's orbit. This figure is greatly exaggerated, since the shift is far too small to be detected without the aid of powerful telescopes.

July

January

your two eyes represent the Earth at opposite sides of its orbit around the Sun and that your finger represents a relatively nearby star, you have the idea of stellar parallax. That is, nearby stars should appear to shift back and forth against the background of more distant stars during the course of each year—*if* the Earth goes around the Sun. (The Greeks actually expected stellar parallax in a slightly different way, since they believed that all stars lay on the same celestial sphere: they expected a shift arising from the fact that at different times of year we would be closer to different parts of the celestial sphere.)

Try as they might, the ancient Greeks were unable to find any evidence of stellar parallax. Today we know that they failed because the stars are too far away for parallax to be detected without the aid of powerful telescopes. For most ancient Greeks, however, it seemed impossible to imagine that the stars could be *that* far away, and they therefore concluded that Aristarchus was wrong and the Earth truly was the center of the universe. Ironically, the successful detection of stellar parallax, first accomplished by the German astronomer Friedrich Bessel in 1838, now constitutes direct proof that the Earth really does go around the Sun. (The first direct proof that the Earth orbits the sun came about a century earlier, when James Bradley detected what is called the aberration of starlight.)

A painting of Tycho Brahe in his naked-eye observatory, from which he made precise measurements of planetary positions among the stars.

Initially, Copernicus had only slightly more luck than Aristarchus with his proposal of a Sun-centered solar system (for reasons we will discuss in Mystery 6). But he inspired a few people to investigate his idea further, and within a century the issue was settled in his favor. The key evidence came from the work of Tycho Brahe, Johannes Kepler, and Galileo.

Tycho Brahe, a Danish astronomer with a nose made from silver and gold (he lost the original at age 20, when he fought a duel with another student over who was the better mathematician), spent more than three decades in the late 1500s compiling the most precise naked-eye measurements ever made of planetary positions among the stars. His observatory on the island of Hven—built with funds from his patron, King Frederick II—was essentially a giant protractor with which he could pinpoint the planets to within 1 minute of arc, which is 1/60 of 1 degree. You can appreciate his observations by again holding your extended finger at arm's length. As seen from your eyes, your finger at this distance covers an angle of about 1 degree—which is sixty times greater than the angles that Tycho could measure. In 1599 he moved, with all of his observational data and a few of his instruments, to the science-oriented court of the Holy Roman Emperor Rudolf II, in Prague, where he hired the young German astronomer Johannes Kepler as his assistant. Upon Tycho's death in 1601, he bequeathed his data to Kepler. (See Episode 3 of Carl Sagan's *Cosmos* series for a moving version of the story of Tycho and Kepler.)

Kepler spent nearly a decade working with Tycho's data before con-

(Top) This photograph, taken through a small telescope, shows Jupiter and the four moons seen by Galileo. Now called the Galilean moons, their names are Io, Europa, Ganymede, and Callisto. (Bottom) In this telescopic photograph, we clearly see, as Galileo saw, that the Milky Way consists of many individual stars that cannot be discerned by the naked eye.

cluding that the planets, including Earth, follow elliptical orbits around the Sun. He summarized planetary motion in three laws (the first two published in 1609 and the third in 1619), with which he could predict planetary positions in a way that precisely matched Tycho's data. Since science generally works by testing how well theoretical predictions match observations, Kepler's theory of planetary motion was an immediate success. However, before it could be fully accepted, there was still the matter of the unobserved stellar parallax, as well as a few other ancient objections to the idea of a Sun-centered solar system. This is where Galileo comes in.

Galileo almost single-handedly refuted every key objection to Kepler's model of the solar system. First, he showed that an object's motion will not change unless it is acted upon by a force. (This idea is now known as *Newton's first law of motion,* even though it was discovered by Galileo.) This defused an objection dating back to Aristotle, who had argued that the Earth could not be going around the Sun because, if it were, airborne objects such as birds and clouds would be left behind as the Earth moved on its way. According to Galileo's new insights into motion, birds and clouds move naturally with the Earth because no outside force is pulling them away from it. Galileo took care of the remaining objections with observations through his telescopes, the first of which he built in 1609. (Credit for inventing the telescope goes to Hans Lippershey in 1608, but Galileo made many improvements to telescope design.) He almost immediately discovered four moons of Jupiter, and his repeated observations of these moons proved beyond any doubt that they orbited Jupiter and not Earth. The idea of the Earth as the center of everything was dead. Many of his other observations added further support to the Copernican idea of a Sun-centered solar system, but one was particularly crucial to defusing the ancient argument about stellar parallax. When he looked into the Milky Way, he found that it was made of countless individual stars, from which he argued that stars were far more numerous and distant than had commonly been thought. Thus he essentially demonstrated that the stars are far enough away to make stellar parallax too small to be noticed by the naked eye, or even with a small telescope like his own.

With Kepler's theory of planets in elliptical orbits matching observations so perfectly, and Galileo's refutation of the remaining objections to a Sun-centered solar system, the evidence in favor of the Copernican idea was overwhelming. By the mid-1600s, the scientific community was nearly unanimous in its acceptance of this new picture of the universe. Even the Church soon stopped arguing about it, though not before convicting Galileo of heresy in 1633. (Galileo was officially restored to good standing in 1992, by Pope John Paul II.) Copernicus had won the day, and our picture of the universe underwent its first dramatic change since the time of Pythagoras, more than two thousand years earlier.

The Copernican revolution ushered in an era of remarkable scientific advancement that continues to this day. We now know not only that the Earth is just one planet going around the Sun but also that our Sun is just one star among billions in the Milky Way galaxy and that our galaxy, in turn, is one among billions of galaxies in the universe. But while we have come a long way in learning the contents of the universe, we still have much to learn about what the universe looks like on large scales.

We know that galaxies often travel in groups. Our own Local Group consists of about thirty galaxies and is held together by gravity. Groups that have many more galaxies are called *clusters,* and the richest galaxy clusters have several thousand member galaxies. We also know that there are structures on even larger scales. Our Local Group lies near the end of a long, finger-shaped collection of many galaxies, groups, and clusters that we call the Local Supercluster (see inside front cover). The Local Supercluster is some 100 million to 150 million light-years long and perhaps 20 million light-years wide. Beyond that, we see many other superclusters, some much longer in extent than our own. But how is it all put together? Are the superclusters strewn randomly about, or are they themselves just small pieces of a much larger structure or structures? Are all galaxies parts of superclusters, or do some galaxies lie in the voids between superclusters? Although evidence is coming in rapidly, we still cannot fully answer these questions. Thus we have our Mystery 8: What does the universe as a whole look like?

This photograph shows the Coma Cluster, a rich cluster containing several thousand galaxies. Almost every object in this picture is a galaxy. Superclusters are much larger still, containing many groups and clusters of galaxies, as well as individual galaxies.

Most people are surprised to learn that we know so little about the large-scale structure of the universe, but it's easy to understand why if you think about how we view the sky. The first problem is that we lack depth perception when we look into the sky, which is why the sky looks like a two-dimensional celestial sphere. When we see an object—whether with the naked eye or through a powerful telescope—the best we can do is give the object's position on the imaginary celestial sphere,

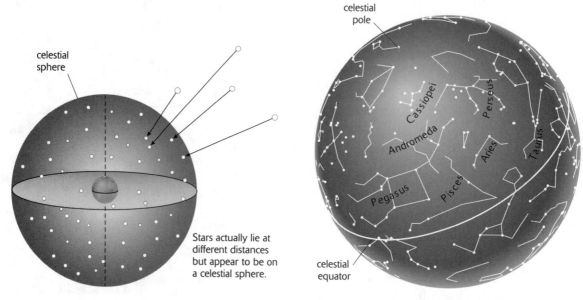

(Left) Because we lack depth perception when we look into space, the sky looks like a two-dimensional celestial sphere surrounding the Earth. (Right) We can map the celestial sphere just as we map the Earth, naming regions by constellation; remember that we imagine the Earth to be in the center of this sphere. This map tells us precisely where to look in the sky for a particular star or galaxy but tells us nothing about the object's distance.

with no indication of the object's distance. Thus, for example, our eyes group the stars into patterns that we call constellations, but these groupings have little to do with reality. Two stars that appear close in the sky may actually lie at vastly different distances. If we want to know how stars and galaxies are arranged in space, we need some way of measuring or at least estimating their distances.

The second problem is that the objects we see are not necessarily representative of what's actually out there. This problem arises because any observing instrument, from the eye to the Hubble Space Telescope, can see only objects whose brightness lies above some instrument-dependent threshold. The brightness of an astronomical object depends both on its distance (since an object will look dimmer if it's farther away) and on its *intrinsic brightness,* or how much light it actually puts out. Thus, even if we knew the distances and hence the three-dimensional arrangement of all the objects we see in the sky, we might still be missing many other objects that lie in between but happen to be intrinsically dim. The stars at night epitomize this problem. You might expect the few thousand of them visible to the naked eye to be a good sample of stars in our Sun's neighborhood of the galaxy, but they are not. The vast majority of stars are much smaller and intrinsically dimmer than our Sun, making them impossible to see with the naked eye even if they are relatively close to us. For example, the nearest star besides the Sun, known as Proxima Centauri (one of three stars in the Alpha Centauri system), cannot be seen by the naked eye.

The challenge of figuring out what the universe looks like comes down to solving these two problems by finding a method of measuring cosmic distances and a way to apply this method even to very dim objects. The extent to which we can solve the latter problem is limited by the current state of telescope and detector technology. While today's best research telescopes have the light-collecting power of more than a million human eyes combined, it is still likely that small, dim galaxies (known as *dwarf galaxies*) are eluding our detection even when they lie just a few million light-years from the Milky Way. There's nothing we can do about this problem but wait for advances in telescope power, so we'll focus our attention on the more subtle problem of measuring cosmic distances.

Measuring cosmic distances essentially means gaining some kind of depth perception for the sky, and for this we return to stellar parallax. If you repeat our earlier experiment of alternately closing your left and right eyes while looking at your finger, you'll notice that you can change the apparent shift by moving your finger closer to or farther from your face—when it's farther, the apparent shift is smaller. In the same way, nearby stars show a greater shift due to stellar parallax than do more distant stars. By adding in a bit of trigonometry and the known Earth–Sun distance, astronomers can use measurements of stellar parallax to calculate distances to stars. Unfortunately, this technique is currently possible only for stars located within a couple of thousand light-years from us, which barely gets us out of our own stellar neighborhood. Beyond that, the shift due to parallax is so small that our current instruments cannot detect it. The most precise parallax measurements currently come from Hipparcos, a European space mission that operated in Earth orbit from 1989 to 1993. Parallax measurements may soon get an even bigger boost from space missions now on the drawing board both at NASA and in Europe. These missions will be capable of measuring much smaller amounts of stellar parallax, and therefore they will be able to measure parallax for much more distant stars. (Some proposals also involve sending spacecraft far beyond Earth's orbit, which would increase the amount of any star's parallax in the same way that widening the distance between your eyes would increase the shift you see when you look at your finger.) With luck, we may soon be able to measure the parallax of objects up to a million or more light-years away.

To measure the distances to objects too far for parallax, we have to rely on indirect techniques. As you might imagine, indirect techniques are prone to great uncertainty, and it takes an enormous amount of work to acquire reliable indirect distance measurements. Indeed, one of the crowning achievements of twentieth-century astronomy was the gradual construction of a reliable cosmic distance scale. The edifice is not yet complete, particularly for objects in the outer reaches of the cosmos, but its foundation is nonetheless solid. Whereas astronomers at the end of the nineteenth century were not even sure whether the universe extended beyond the Milky Way, we can now estimate the distances to the farthest galaxies with confidence that our measurements lie within about 20 percent of the true values.

The historical route to building the cosmic distance scale was long and tortuous, but the structure is easy to understand with hindsight. It relies on the idea of what astronomers call *standard candles,* a term meant to suggest light sources of known intrinsic brightness. For example, a 100-watt lightbulb is a standard candle, because we can safely assume that all such lightbulbs put out the same amount of light. If we know that we are looking at a 100-watt lightbulb, we can determine its distance from us by how bright it looks to us. The same idea works with stars. For example, we can safely assume that any star of the same type as our Sun has about the same intrinsic brightness as our Sun, and therefore we can determine its distance by its apparent brightness in our sky. The only trick is that we need a way to classify stars by type so that, for example, we know whether a star is of the same type as the Sun or of a type that is intrinsically dimmer or brighter.

The trick of classifying stars was discovered early in the twentieth century, largely by a group of women astronomers working at the Harvard College Observatory. Most of them had studied physics or astronomy at women's colleges such as Wellesley and Radcliffe but were unable to continue their formal studies because institutions offering advanced degrees, such as Harvard, would not admit women. Fortunately for science, the director of the Harvard Observatory, Edward Pickering, had a problem: by the late 1800s, his observatory was collecting far more data than he or his male assistants had time to study. Pickering therefore began hiring women for positions that he called "computers," since their jobs called for making measurements and computations with spectra collected

Women astronomers pose with Edward Pickering at the Harvard College Observatory in 1913. Annie Jump Cannon is fifth from the left in the back row.

at the observatory. Perhaps to the chagrin of the leaders of the male-dominated institution, the women had the audacity to make their own scientific discoveries as they sifted through the data, thereby revolutionizing astronomy and forcing open the doors of science to many other women. Our modern scheme for classifying stars was developed by Annie Jump Cannon. During her career, Cannon personally classified more than four hundred thousand stars.

To understand how stellar classification works, think of how a prism disperses sunlight into the spectrum that we know as the rainbow. Starlight can also be dispersed into a spectrum, using instruments called spectrographs. When examined with the detail afforded by precision spectrographs, stellar spectra are filled with spectral lines that can tell us the chemical composition, temperature, pressure, and even rotation rate of the stars we study. Most important to the cosmic distance scale, the pattern of the spectral lines tells us the star's type (its "spectral type"), from which we can infer its intrinsic brightness. Once we know a star's intrinsic brightness, we can determine its distance by how bright it looks in the sky.

Because most stars outside our own galaxy are too dim to see even with large telescopes, stellar classification takes us only partway up the cosmic distance scale. Fortunately, astronomers have also learned to use intrinsically brighter objects as standard candles—objects such as rare but extremely bright variable stars known as *Cepheids* (which we'll discuss in Mystery 7), the exploding stars known as supernovae, and sometimes even entire galaxies. Together, these techniques have helped us develop a fairly simple way of measuring the distance to almost any galaxy—a method that relies on something called *Hubble's law.*

Hubble's law gets its name from Edwin Hubble, who discovered the astonishing fact that the universe is expanding. More specifically, Hubble's law tells us this:

The farther away a galaxy is located, the faster it is moving away from us.

Or, conversely:

The faster a galaxy is moving away from us, the farther away it is.

Upon first hearing this law, you might be tempted to conclude that our Local Group—which is held together by gravity—suffers from a cosmic case of chicken pox that leads other galaxies to race away from us. But it's easy to see that universal expansion offers a more natural explanation by thinking about a raisin cake baking in an oven. Imagine that you make a raisin cake and the distance between adjacent raisins is 1 centimeter. You place the cake in the oven, where it expands as it bakes. After an hour, you remove the cake, which has expanded so that the distance between adjacent raisins is 3 centimeters (see diagrams on facing page).

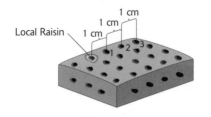

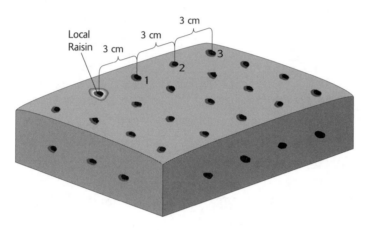

The raisins in an expanding raisin cake separate much like clusters of galaxies in the expanding universe.

Pick any raisin (it doesn't matter which one), call it the Local Raisin, and identify it in the pictures of the cake both before and after baking. Before baking, the nearest raisin to the right, or Raisin 1, is 1 centimeter away from the Local Raisin, Raisin 2 is 2 centimeters away, and Raisin 3 is 3 centimeters away. After baking, Raisin 1 is 3 centimeters away from the Local Raisin, Raisin 2 is 6 centimeters away, and Raisin 3 is 9 centimeters away. Nothing should seem surprising so far, because the new distances simply reflect the fact that the cake has expanded uniformly everywhere.

Now suppose you live *inside* the Local Raisin. From your vantage point, the other raisins appear to move away from you as the cake expands. For example, Raisin 1 is 1 centimeter away before baking and 3 centimeters away after baking. It therefore appears to move 2 centimeters during the hour, so you'll see it moving away from you at a speed of 2 centimeters per hour. Raisin 2 is 2 centimeters away before baking and 6 centimeters away afterward; thus it appears to move 4 centimeters during the hour of baking, giving it a speed of 4 centimeters per hour away from you. The following table lists the distances of various raisins before and after baking, along with their speeds as seen from the Local Raisin.

	Distances and Speeds of Other Raisins as Seen from the Local Raisin		
Raisin Number	Distance Before Baking	Distance After Baking (one hour later)	Speed
1	1 cm	3 cm	2 cm/hr
2	2 cm	6 cm	4 cm/hr
3	3 cm	9 cm	6 cm/hr
4	4 cm	12 cm	8 cm/hr
⋮	⋮	⋮	⋮
10	10 cm	30 cm	20 cm/hr
⋮	⋮	⋮	⋮

Note that although the entire cake expanded uniformly, from the vantage point of the Local Raisin, each subsequent raisin moved away at increasingly faster speeds. Furthermore, because selection of the Local Raisin was arbitrary, you will find the same results no matter which raisin you choose to represent your Local Raisin. Thus, if you lived in an expanding raisin cake, you would discover the following analogue to Hubble's law: The farther away a raisin is located, the faster it is moving away from you.

If you now imagine the Local Raisin to represent our Local Group of galaxies, and the other raisins to represent more distant galaxies, you have a basic picture of the expansion of the universe. That is, space itself is growing, and Hubble's law—that galaxies farther away from us are moving away from us faster—arises because the galaxies are carried along with this expansion like raisins in an expanding cake. (Individual galaxies and groups of galaxies are not expanding, because they are held together by gravity.)

In quantitative terms, Hubble's law gives a direct proportionality between a galaxy's distance and its speed. The proportionality is not quite perfect, because in addition to being carried along with the universal expansion, galaxies can also be moving in response to gravitational pulls. However, except for relatively nearby galaxies, these other motions are small compared with what we see because of expansion, so Hubble's law can give us an excellent estimate of a galaxy's distance if we first measure its speed.

Measuring a galaxy's speed turns out to be fairly easy if we study its spectrum. You're probably familiar with a phenomenon called the *Doppler effect,* which gives an approaching car or train a higher-pitched sound than that of a receding car or train. The Doppler effect also works for light, shifting the light from approaching objects to higher frequency (shorter wavelength) and the light of receding objects to lower frequency (longer wavelength). Because red light has a lower frequency than blue light, astronomers say that the light of a receding object shows a *redshift.* In other words, the spectrum of a distant galaxy will have a redshift that moves all of its light, including its spectral lines (which serve as the reference marks), in the direction of the red end of the rainbow. The larger the redshift, the faster the galaxy is receding.

Thus, by Hubble's law, the more distant the galaxy, the greater the redshift. In principle, then, we can use Hubble's law to determine a galaxy's distance directly from its redshift. In practice, the fact that our cosmic distance scale is still uncertain by up to 20 percent means a corresponding uncertainty in our calibration of Hubble's law. As a result, astronomers generally prefer to talk about distant galaxies in terms of their redshifts rather than

These three schematic spectra illustrate the idea of redshift. The top spectrum shows the positions of spectral lines produced by hydrogen gas, as measured in laboratories on Earth. The middle spectrum shows the same lines in light from a moderately distant galaxy; note that they show a *redshift* toward the red side of the spectrum. The bottom spectrum shows the same lines from an even more distant galaxy, which therefore has an even greater redshift.

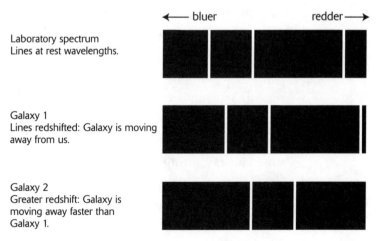

← bluer   redder →

Laboratory spectrum
Lines at rest wavelengths.

Galaxy 1
Lines redshifted: Galaxy is moving away from us.

Galaxy 2
Greater redshift: Galaxy is moving away faster than Galaxy 1.

their uncertain distances. For example, you may hear an astronomer say that a particular galaxy has a redshift of 0.5, while a more distant galaxy has a redshift of 4.2. Journalists usually ask astronomers to convert these distances into estimates in light-years, but remember that these distance estimates may be off by up to 20 percent even though the redshifts are quite precise.

By now, you can probably see how we could make a complete, three-dimensional map of the universe. We simply need to measure redshifts for every galaxy out there. With each galaxy's redshift serving as a proxy for its distance, and with its location among the constellations telling us its direction, we could place each galaxy correctly in a three-dimensional model of the universe. Unfortunately, this task is much more difficult in practice than in theory.

Large telescopes can photograph only a tiny portion of the sky at a time, and taking a picture that shows very faint galaxies can require hours or even days of exposure time. Moreover, a photograph doesn't tell us redshifts; for that, we need to go back and obtain individual spectra of the galaxies in the picture. Given that a single photograph may show thousands of galaxies, it can be extremely time-consuming to get three-dimensional data for the galaxies in even a tiny piece of the sky.

The result is that, here at the dawn of the third millennium, astronomers have succeeded in mapping the locations of only a very small fraction of the estimated 100 billion galaxies in the observable universe. The two wedges in the picture on the facing page show some ten thousand galaxies that have been mapped in two thin slices of the sky. These data, produced by a team at the Harvard-Smithsonian Center for Astrophysics led by astronomers Margaret Geller and John Huchra, took nearly fifteen years to collect. You'll understand what you're seeing if you imagine holding each wedge with its vertex at your nose and extending it into space. Our Milky Way galaxy lies at the central vertex, and each dot represents another galaxy. The farther the dot is from the center, the farther away the galaxy lies. Thus each wedge shows how galaxies are arranged on a thin triangle extending

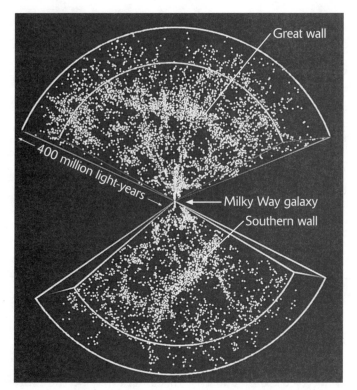

Great wall

400 million light-years

Milky Way galaxy
Southern wall

These wedges show the approximate distribution of a few thousand bright galaxies that lie within about 400 million light-years of the Milky Way. Each dot represents a single galaxy. The upper wedge is a thin slice of the sky looking to the north, and the lower wedge looks to the south.

from here out into the universe. The radius of these maps—that is, the distance from the vertex to the far edge—is about 400 million light-years.

If you study the wedges carefully, you'll see that the galaxies are not strewn about randomly. Instead, they appear to be arranged in huge chains or sheets that stretch across tens of millions of light-years of space. The overall appearance is much like a gigantic network of soap bubbles, with galaxies found along the bubble walls. In between are giant voids, which seem to hold few if any galaxies. Some of the structures in these small pieces of the sky are amazingly large. The so-called Great Wall of galaxies, which stretches across the center of the upper wedge, measures at least 500 million light-years from end to end. Because it runs off the right side of the wedge, it could be much larger still. In fact, with a little imagination, you might envision it being connected to the similar Southern Wall, which cuts a swath through the lower wedge.

Because we've mapped so little of the universe in three dimensions, larger structures may still be undiscovered. These structures may dwarf even the largest-known superclusters. To use an old cliché, the twentieth-century emphasis on studying individual galaxies and clusters of galaxies may have left us missing the forest for the trees. Learning more about the forest is vitally important to understanding the ecology of the universe as a whole.

Several groups of astronomers have therefore embarked on intensive efforts to map the universe better. These efforts are often called redshift surveys, since they generally involve measuring the redshifts of many thousands of galaxies. The most ambitious effort involves a collaboration of more than a hundred scientists at nearly a dozen institutions, and goes by the name of the Sloan Digital Sky Survey (because major funding came from the Alfred P. Sloan Foundation). The Sloan survey uses a 2.5-meter telescope at the Apache Point Observatory in New Mexico (see Color Plate 5). Its camera may be the most complex ever built, with thirty small (2-inch square) electronic detectors called CCDs (for Charge-Coupled Device) containing more than 4 million pixels (picture elements) each. A single night's observing can produce up to 200 gigabytes of data—enough that if stored as words on a page they would fill a library of nearly a half million books. Over its planned ten years of operation, the Sloan survey will produce about 15 terabytes of data, rivaling the total information content of the Library of Congress. The telescope achieved first light (its first real images) in May 1998. Now, after a period for calibrating the telescope and cameras, the real survey is under way.

The Sloan survey aims to obtain high-quality photographs of about one-quarter of the entire sky. Based on the sensitivity of the telescope and camera, these images should capture the light of more than 70 million stars and 50 million galaxies. Computer data processing will be used to compile all these data into a catalog giving the precise positions and brightnesses of every one of these objects. It will be by far the most complete survey ever undertaken of the contents of the universe.

For the purposes of uncovering large-scale structure, the key to the Sloan survey lies in its plans for measuring the redshifts of about 1 million galaxies. The redshifts will be measured with a technological marvel: a spectrograph that uses 640 optical-fiber cables to simultaneously obtain spectra of 640 objects. The painstaking procedure works essentially like this: After obtaining a photograph of a particular region of the sky, scientists use a computer to help identify the 640 brightest galaxies. A robotic system then drills 640 holes into a metal plate at positions corresponding to the positions of the galaxies in the photograph. Working by hand, astronomers plug the 640 optical-fiber cables into the holes. The entire setup is then attached to the telescope, which is then pointed to the same spot in the sky. This time, the light from the 640 galaxies passes into the optical fibers, producing spectra from which the individual redshifts of the galaxies can be measured. The drilling and plugging will usually happen during the day, so that on a good night astronomers may be able to obtain several of these 640-spectra data sets.

We gain an enticing hint of what the Sloan survey will find by looking at what earlier surveys have shown in two dimensions. The picture on the next page shows the distribution of over 3 million galaxies spread across about 15 percent of the sky, which is a region as large as several major constellations put together. In other words, this picture is roughly what you would see in the night sky if we could somehow turn off the stars and give your eyes enough sensitivity to see the faint light of distant galaxies. Even without knowing the distances to the individual galaxies, it is clear that the universe is filled with some kind of weblike structure. By recording more than ten times as many galaxies in two dimensions, and adding the third dimension with the redshifts for a million galaxies, the Sloan survey should begin to give us our first accurate picture of how this intricate structure is woven into the universe. The general picture should be available as early as 2005, though the enormous amount of data ensures that important discoveries will continue to be made with the Sloan data for many years after that.

This picture is a composite that shows the distribution of more than 3 million galaxies spread across some 15 percent of the sky. Although it is only a two-dimensional picture (because it does not tell us distances to the galaxies), we see hints of intricate structure. (Black rectangles are regions for which there are no data.)

Learning what the universe looks like will undoubtedly tell us many important facts about its history. Indeed, the enormous structures already discovered came as a big surprise when astronomers first began to notice them in the 1980s. Given that the universe is no more than about 16 billion years old, simple calculations show that gravity could not possibly have made such huge structures from scratch in the time since the Big Bang. Instead, the structure must have grown from seeds that were somehow imprinted on the universe in its very early history, thereby leaving gravity the easier task of collecting matter around this existing imprint. By analogy, we can think of gravity as the construction crew putting up the floors, walls, and ceilings on a building that has already been framed. Just as we could not expect to understand the building's architecture without knowing the structure of the framing, we cannot expect to understand the architecture of the universe until we have a much better picture of what its framing actually looks like.

Ten thousand years ago, our ancestors imagined the universe as little more than a flat, dome-covered Earth. One thousand years ago, our ancestors envisioned a round Earth in the center of a relatively small universe, surrounded by the celestial sphere. One hundred years ago, we knew that our solar system was part of the larger structure of the Milky Way galaxy, but no one yet knew whether there were other galaxies beyond the Milky Way. Now, in the first decade of the third millennium, we should finally learn what the observable universe really looks like. Will this picture allow us to deduce anything about the architecture and history of the universe? Might it reveal structures so large that they go beyond the bounds of the observable universe, making it impossible for us to see their full extent? Whatever the case, answers should soon be forthcoming.

The infinitude of creation is great enough
to make a world, or a Milky Way
of worlds, look in comparison with it
what a flower or an insect does
in comparison with the Earth.

IMMANUEL KANT (1724–1804)

How Do Galaxies Evolve?

Galaxies serve as giant recycling plants, wherein the elements manufactured by stars are recycled into new generations of stars and planets. Thus, if we hope to understand the complete story of our cosmic origins, we must first understand how galaxies were born and how they change with time. In this mystery, we will explore our current understanding of galaxy evolution, along with some of the key questions that remain unanswered. We will see how coming observations may help us finally understand the lives of galaxies, and how the quest to understand the birth of galaxies takes us all the way back to the earliest moments of the universe.

As Carl Sagan said, we are "star stuff." The early universe contained only the elements hydrogen and helium (and trace amounts of lithium), and these two elements comprise some 98 percent of the universe's chemical composition today. But we are made primarily of heavier elements, such as carbon, nitrogen, oxygen, and iron—elements forged by nuclear fusion in massive stars that lived and died before our solar system was born. The violent explosions, or *supernovae,* that ended the lives of such stars scattered their storehouses of manufactured elements into space.

But if that were the end of the story, we wouldn't be here. Something had to prevent the scattered elements from simply dispersing into the expanding universe, and for that we need galaxies. Within a large galaxy, gravity keeps the scattered elements contained, allowing them to mix with other interstellar gas and dust. Then gravity can slowly work its magic, collecting this material into vast clouds that give birth to new generations of stars. About 4.6 billion years ago, one such cloud gave birth to our Sun and its system of planets. In essence, galaxies are the cosmic recycling plants that allow elements manufactured in one generation of stars to be recycled into subsequent generations. In that sense, we are "galaxy stuff" as well as "star stuff," because both stars and galaxies are requirements for our existence.

Today we seem to understand stellar life cycles quite well, and theoretical models of element formation in stars match our observations of the relative abundances of elements in the cosmos. But because we are galaxy stuff, we also need to understand the lives of galaxies if we hope to know the complete story of our cosmic origins. Unfortunately, we still know very little about how galaxies are born and how they live, which brings us to our Mystery 7—the mystery of how galaxies evolve with time.

One of the most remarkable things about the mystery of galaxies is how new it is. Given that the observable universe is filled with 100 billion or more galaxies and that they appear to be one of the most fundamental organizational units in the universe, you might think that questions about galaxies would go far back into astronomical history. In fact, only a century ago no one was even sure whether galaxies existed beyond the Milky Way.

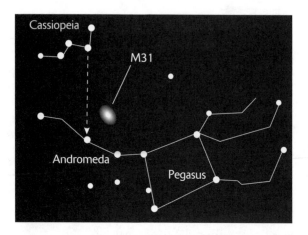

This diagram shows the location of M31, the Great Galaxy in Andromeda, among the constellations. From a dark site, you can see it with the naked eye. It appears as a small, fuzzy patch, but it is the combined light of more than 100 billion stars. Color Plate 1 shows a telescopic photograph of M31.

The problem was not in seeing galaxies, but in figuring out what they were. You can even see one large galaxy with your naked eye—M31, or the Great Galaxy in Andromeda—if you look from a dark site. Two smaller galaxies, the Large and Small Magellanic Clouds, are also visible to the naked eye, but only from equatorial latitudes and the Southern Hemisphere. Today, it can be thrilling to look at M31—at least once you realize that you are seeing the combined light of more than 100 billion stars and that this light has traveled through space for 2.5 million years before striking your retina. But without our modern knowledge, M31 looks like nothing more than a fuzzy little cloud. Binoculars and small telescopes don't change this basic appearance.

In the eighteenth and nineteenth centuries, astronomers discovered many such fuzzy patches in the sky. Because they looked like small clouds, or nebulae, and many had a noticeably spiral shape, they were often called *spiral nebulae.* The telescopes of the time were not up to the task of seeing individual stars in these patches, so their true nature remained unclear. Many astronomers assumed they were spinning clouds of gas interspersed among the stars of the Milky Way, and some thought they might be the birthplaces of solar systems like our own. In 1755, the German philosopher Immanuel Kant guessed correctly that the spiral nebulae were vast communities of stars, isolated from the Milky Way and one another by chasms of empty space. The quote at the beginning of this mystery captures his sense of wonder regarding these stellar communities, which later writers dubbed "island universes."

Unfortunately for proponents of the island-universe theory, guessing right is never good enough in science. Science demands hard evidence, and little of relevance to the nature of the spiral nebulae was forthcoming for a hundred and fifty years after Kant's original proposal. During the first two decades of the twentieth century, advances in astronomical technology and data collection began to arm astronomers with facts about the spiral nebulae, and for a while these facts fueled both interpretations. The issue became so contentious that in 1920 the National Academy of Sciences sponsored what has since become known in astronomical circles as The Great Debate.

The Great Debate took place in Washington, D.C., on April 26, 1920; the official topic was "The Scale of the Universe." Harlow Shapley, a rising young astronomer who would become director of the Harvard Observatory the following year, took the view that the spiral nebulae were gas clouds internal to the Milky Way and therefore that the universe was composed only of our own, single galaxy. Heber Curtis, of the Lick Observatory, argued in favor of Kant's island universes, which made the universe immense and possibly infinite. Most observers at the time scored the debate in favor of Curtis and the island universes, but the case was certainly not closed.

Closure came in 1924, thanks to the work of Edwin Powell Hubble. Hubble was a towering presence in twentieth-century astronomy, both literally and figuratively. He excelled in athletics, competing in basketball, track, and heavyweight boxing during his undergraduate years at the University of Chicago. He was a particularly skilled boxer and once fought an exhibition match against the reigning world light-heavyweight champion. Offered a chance to box professionally upon his graduation in 1910, he instead went off to Oxford as a Rhodes

Edwin Hubble posing with the 48-inch telescope on Mount Palomar.

Scholar. There he spent three years studying law. He returned home to Louisville, Kentucky, in 1913 and took a job as a high school teacher across the Ohio River in New Albany, Indiana. He taught physics, mathematics, and Spanish, and coached the boys' basketball team to an undefeated regular season and third place in the Indiana state championship. The tall and dashing Hubble proved so popular among the students that the Class of 1914 dedicated its yearbook to him. But Hubble had already decided that his true passion lay with astronomy, so he left Kentucky and entered a graduate program at the University of Chicago's Yerkes Observatory.

Hubble completed his doctoral work in 1917 and was invited to join the staff at the Mount Wilson Observatory in Pasadena, California. This was a prestigious invitation, as work was just finishing on Mount Wilson's jewel, a 100-inch telescope that would be the world's largest for the next thirty years. But in April 1917, the United States entered the war against Germany, and instead of accepting the invitation Hubble volunteered for service. He telegraphed the observatory, saying "Regret cannot accept your invitation. Am off to the war." He served in France, rising to the rank of major before returning to the United States, where he finally joined the staff at Mount Wilson in 1919. There he stayed until 1942, when he again volunteered for service and became the head of ballistics at the Aberdeen Proving Ground in Maryland. For his World War II service, he was awarded the Medal of Merit, the highest civilian award granted by the president, in 1946.

As he settled in at Mount Wilson in the early 1920s, Hubble turned his attention to the study of the spiral nebulae. With the 100-inch telescope, he could see what looked like individual stars in the Andromeda galaxy, strongly suggesting that Kant's idea was right. But his breakthrough came when he discovered that some of these stars were dimming and brightening with a regular period. This marked them as stars of a special type known as *Cepheid variables* (so named because the prototype star appears in the constellation Cepheus).

A decade earlier, the Harvard astronomer Henrietta Leavitt had discovered that Cepheid variable stars exhibit a peculiar relationship between intrinsic brightness and period: the longer the period between one peak of brightness and the next, the brighter the star. Today we know why. Cepheids pulsate physically; that is, they periodically grow

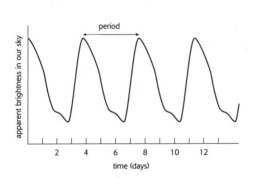
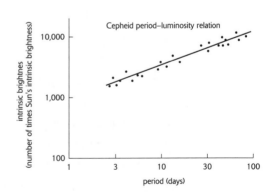

(Left) This graph shows how the apparent brightness of a Cepheid variable changes over a regular period. In this example, the period is about 4 days. (Right) Cepheids follow a "period–luminosity relation," in which their periods tell us their intrinsic brightness. For example, this graph shows that a Cepheid with a period of 4 days is more than 1,000 times brighter than our Sun.

larger and smaller in radius, with corresponding changes in brightness. There is also a direct correlation between the mass of a Cepheid and its intrinsic brightness, and a more massive star takes longer to complete its pulsation. Hence the brighter Cepheids have longer periods.

By virtue of their periods alone, the Cepheids in Andromeda were announcing their intrinsic brightnesses. This meant that Hubble could use them as standard candles (see Mystery 8). Because their periods told him their intrinsic brightnesses, he could calculate their distances by measuring how bright they appeared in the sky. The calculations immediately showed that the Cepheids in Andromeda lay far beyond the bounds of the Milky Way. The debate was over; the spiral nebulae really are distinct galaxies of stars.

Today, with the aptly named Hubble Space Telescope (see Color Plate 6), we can see galaxies scattered throughout the universe. Take a look at the remarkable image known as the Hubble Deep Field, which is shown in Color Plate 7. Nearly every small dot in this picture is an entire galaxy of stars. The Hubble Space Telescope required ten days of exposure time to acquire this image, which shows only a tiny piece of the sky in the direction of the Big Dipper—a piece so small that you could cover it with a grain of sand held at arm's length. Some of the light captured in this photograph is 10 billion times too faint to be seen by the naked eye.

The Hubble Deep Field is beautiful, but it also shows why the origin and structure of galaxies remain fundamentally mysterious. Notice that while many galaxies have the characteristic spiral shape, others are bloblike *elliptical galaxies,* and a few others have various irregular shapes. The galaxies also appear to come in a variety of sizes and colors. Moreover, as we'll discuss shortly, the Hubble Deep Field shows that galaxies had formed within a couple of billion years after the Big Bang. If the universe began as a uniform conglomeration of gas, gravity alone could not have formed galaxies that quickly. Instead, gravity must have built galaxies around seeds planted during the Big Bang itself. (See the discussion of gravity and large-scale structure in Mystery 8.) The Hubble Deep Field tells us that we have much to explain if we want to un-

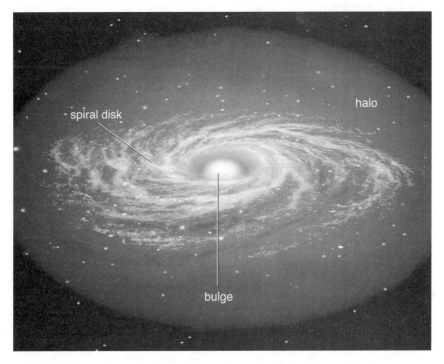

This diagram shows how a spiral galaxy's disk is embedded within a much larger, spherical halo.

derstand how galaxies developed from their seeds into the variety that we see today.

The birth of galaxies must have involved the action of gravity collecting gas around the seeds. If we assume that our own Milky Way formed in this way, we can learn a lot about galaxy formation by studying our own galaxy.

The Milky Way is a spiral galaxy, which means that it looks much like a flat pinwheel. But this pinwheel, more formally known as the galaxy's *disk*—is not the entire galaxy. The disk is embedded within a much larger, spherical volume, which astronomers call the *halo* of the galaxy. Although the halo is far larger than the disk, it is nearly invisible because almost all of the galaxy's bright stars are in the disk. Many of the stars in the halo reside in roughly spherical collections of a few hundred thousand stars known as *globular clusters.*

There are plenty of mysteries about halo stars, but for our current purposes we need to know only that they differ from the stars of the disk in three important ways: (1) halo stars orbit the center of the galaxy with many different and seemingly random orientations, whereas the stars in the disk all orbit in the same direction and the same plane; (2) the stars in the halo are all very old—10 billion years or more—whereas the disk has stars of all ages; and (3) the halo stars are more nearly pure hydrogen and helium than the disk stars (halo stars are typically 99.9 percent hydrogen and helium, as opposed to about 98 percent for disk stars).

The fact that the halo stars are old and deficient in elements besides hydrogen and helium tells us that they were born early in the galaxy's history, and their random orbits tell us that this early period occurred before the galaxy had a chance to organize itself into a spinning disk. These facts give us a simple outline for the basic history of the Milky Way:

- The galaxy began as a blobby cloud of hydrogen and helium gas, known as a *protogalactic cloud*. (In fact, evidence suggests that this protogalactic cloud probably formed from the merger of several smaller clouds.) The gas slowly orbited the center of the cloud, but with no overall organized motion. Different blobs of gas orbited on different, randomly oriented orbits around the center.

- Over time, gravity compacted this cloud of gas, and as it shrank, it flattened into a spinning disk. The cloud spun faster and faster as it shrank, for the same reason that an ice skater spins faster when she pulls in her arms (a phenomenon known in physics as "conservation of angular momentum"). It flattened because any gas particles that were not orbiting in the plane of rotation collided with others that were, which gradually forced them into the plane.

This simple scenario explains the differences between halo and disk stars. The halo stars formed when the galaxy was still bloblike, which is why they have randomly oriented orbits. (Once a star forms, it is so much denser than any surrounding gas that it generally retains its original orbit, no matter what happens to the gas.) They are all old because the leftover gas fell into the disk, leaving nothing with which to produce new generations of stars in the halo. They contain only a minuscule proportion of heavy elements because they were born before these elements had been created in greater quantities by multiple generations of stars. In contrast, disk stars belong to the flat, spinning portion of the galaxy, in which matter is continually recycled between stars and interstellar gas, making possible the creation of new generations of galaxy stuff—such as planets and people.

Studies of other spiral galaxies reveal similar distinctions between their relatively few halo stars and their many disk stars. Thus this basic scenario for the Milky Way's history probably applies to most spiral galaxies.

Studies of elliptical galaxies show that they also fit into this general picture, with one important modification. The motions of stars in most elliptical galaxies look remarkably like those of stars in the Milky Way's halo. Like halo stars, stars in elliptical galaxies have randomly inclined orbits around the galactic centers, which explains why elliptical galaxies have the same basic shape as the halos of spiral galaxies. In essence, then, elliptical galaxies look like spiral galaxies without the spiral disk. In that case, we can guess that elliptical galaxies also began as protogalactic clouds but that they somehow ran out of gas with which to make a disk—perhaps because they used all their gas to make stars before a disk had a chance to form or because any gaseous disk that once formed was later destroyed.

Scientists today can test some of these ideas about galaxy formation with the aid of supercomputer simulations. The scientists write programs that incorporate the known laws of physics and then run these programs on various sets of starting parameters to see whether they can create computer galaxies that look like real galaxies. These models show that our scenario for galaxy development is plausible—as long as one of the starting parameters is a universe filled with seeds around which gravity can act. The models start with the idea that shortly after the Big Bang the entire universe was expanding uniformly. But the seeds, which in essence are regions of slightly enhanced density, have stronger gravity than surrounding regions. Gravity thereby slows the universal expansion around the seeds, and within about a billion years after the Big Bang, the expansion in these small regions halts altogether. Once the expansion is halted around a seed, gravity can collect the matter in the area into a shrinking protogalactic cloud.

Of course, plausible does not mean true, so it would be nice to be able to get more direct evidence about how galaxies evolve. But how? We might wish for a time machine that would allow us to view the universe at various times in the past. Then we could see what galaxies looked like a billion years ago, or 5 billion years ago, or all the way back to the first billion years after the Big Bang. With such a time machine, we could in principle resolve all the mysteries of galaxy evolution. No one yet knows if time travel is possible, but you may be surprised to learn that we already have "time machines" that allow us to see into the past. We call them telescopes.

The idea of a telescope as a time machine is really quite natural. Light takes time to travel through space, which is why we measure astronomical distances in light-years. (Recall that a light-year is the distance that light travels in one year, which is about 10 trillion kilometers or 6 trillion miles.) Because it takes time for light to travel through space, *the farther away we look in distance, the further back we look in time.*

If we look at a galaxy that lies 10 million light-years away, we are seeing it as it was when the light left it 10 million years ago. If we observe a cluster of galaxies that lies 1 billion light-years away, we are seeing the cluster as it was 1 billion years ago. Our telescopes allow us to peer directly into the past, simply by giving us the power to see distant objects.

In fact, it's actually more sensible to think of faraway galaxies in terms of what astronomers call their *lookback times* than in terms of distance. By "lookback time," we mean the amount of time that the light has taken to reach us. For example, when we say that a galaxy has a lookback time of 1 billion years, we mean that its light took 1 billion years to reach us and hence we are seeing it as it was a billion years ago. In contrast, saying that this galaxy is "a billion light-years away" is ambiguous, because cosmic distances change as the universe expands. The distance statement is unclear as to whether it means the distance to the galaxy now, the distance at the time the light left (a billion years ago), or some other distance in between. When dealing with very distant objects, astronomers therefore prefer the concrete idea of lookback time to the ambiguous idea of distance.

The important thing to remember about lookback time is that it is greater for more distant galaxies, which means we are seeing such galaxies as they were when the universe was younger. To see how this works, suppose the universe is 12 billion years old. If we observe a galaxy with a lookback time of a billion years, we are seeing it as it was when the universe was a billion years younger than today, or 11 billion years old. Similarly, we see a galaxy with a lookback time of 6 billion years as it was when the universe was only 6 billion years old, or half its current age. A galaxy with a lookback time of 11 billion years appears as it did when the universe was a mere billion years old.

While we're at it, note that if the universe is 12 billion years old, we could not possibly see anything with a lookback time of more than 12 billion years, because that would mean seeing back to a time before the universe existed. Hence, for a 12-billion-year-old universe, a lookback time of 12 billion years represents a *cosmological horizon* beyond which we cannot see. This cosmological horizon defines the boundaries of our observable universe. Notice that it is not a physical boundary but a boundary in time. With each passing year, the lookback time to the cosmological horizon increases by a year, thereby enlarging (very slightly!) our observable universe.

The fact that larger lookback times mean seeing into the universe when it was younger gives us a remarkable ability: we can essentially create a family album of what galaxies looked like at different stages in the history of the universe by putting together photographs of galaxies with different lookback times. The only hard part is obtaining photographs of galaxies with very large lookback times. Because these galaxies are extremely far away, their light is much dimmer than that of nearby galaxies. Moreover, if we try to look back far enough in time, we look to an era before stars began to shine, and hence when galaxies were still dark. At present, we have a reasonable start to a family album of galaxies going back to a time just 2 or 3 billion years after the Big Bang, but we have yet to obtain clear photographs of galaxies in their earliest stages, or as protogalactic clouds.

Despite its limitations, our available family album of galaxies has revealed plenty of surprises. Galaxies are not quite the island universes that Kant imagined; rather, galaxies frequently interact and sometimes collide head-on. A galaxy collision unfolds over a billion years or more, so we cannot actually watch these spectacular cosmic events. But by seeing galaxies in various stages of collision and modeling the process with supercomputers, we have learned that interactions play a major role in the lives of galaxies.

Since stars within the galaxies are typically separated from one another by light-years of space, stellar collisions during a galactic collision are vanishingly rare. But the orbits of stars are changed by the gravitational effects of the collision, and sometimes two galaxies will effectively merge. In addition, the gas within the galaxies can be severely disrupted; in some cases gas may be stripped out of the colliding galaxies, while in other cases huge clouds of gas may begin to collapse, leading to a burst of rapid star formation. In at least a few cases, galactic collisions almost certainly explain the "somehow" that rids elliptical galaxies of any gaseous disk. Models show that when two spiral galaxies collide and merge, the result can be a diskless—and hence elliptical—galaxy.

Perhaps even more astonishing than these tremendous cosmic collisions are the turbulent youths of many galaxies. The most powerful objects in the universe are the so-called *quasars,* which got their name as shorthand for "quasi-stellar radio objects." As the name suggests,

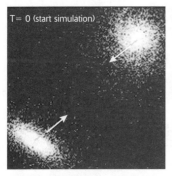

T= 0 (start simulation)

Two simulated spiral galaxies approach each other on a collision course.

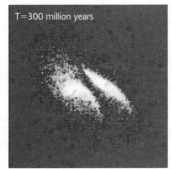

T=300 million years

The first encounter disrupts the two galaxies and sends them into orbit around each other.

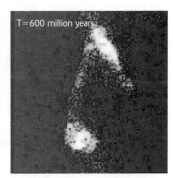

T=600 million years

Gravitational forces between the galaxies tear out long streamers of stars.

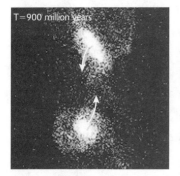

T=900 million years

Because the first disruptive encounter saps kinetic energy from the galaxies, they cannot escape each other.

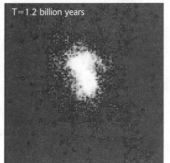

T=1.2 billion years

The second encounter is more direct than the first; the galaxies collide head-on and begin to merge.

T=1.5 billion years

The single galaxy resulting from the collision and merger is an elliptical galaxy surrounded by debris.

Several stages in a supercomputer simulation of a collision in which two spiral galaxies merge to form a single elliptical galaxy. The whole sequence spans about 1.5 billion years.

the first ones discovered were found by virtue of powerful radio emission. When optical telescopes were trained on them, they looked much like ordinary stars, except that their spectra showed huge redshifts. Quasars were a great mystery for a couple of decades after their discovery in the early 1960s. Their redshifts suggested that they were very distant objects being carried along with the expansion of the universe, but some astronomers argued that the large redshifts arose from some other cause. That debate has been all but settled by improved observations, which have revealed the environs in which quasars live. We now know that quasars do not really represent a new type of object, but a stage of life for the central regions of many young galaxies. More specifically, a quasar is a galactic center that is pouring out an enormous quantity of energy from a very small space.

The brightest quasars shine more powerfully than a thousand galaxies the size of the Milky Way, yet detailed studies show that all this power comes from a region not much larger than our solar system. The question of what could generate such tremendous power was a major mystery of the late twentieth century, but most astronomers now consider this mystery solved at least in general, if not in the details. Quasars are powered by the energy released when matter in the center of a galaxy falls into a central, *supermassive black hole* (see Color Plate 8).

You've probably heard about black holes, into which matter can enter but never return. According to the modern theory of quasars, the mass of the central black hole may be a billion or more times the mass of our Sun. It turns out that infall of matter into a black hole is an awesomely efficient way to generate energy. Recall that, by virtue of its mass, all matter contains "mass-energy" in the amount $E = mc^2$ (where E is the energy it contains, m is its mass, and c is the speed of light). Nuclear fusion, which we usually think of as being a very efficient way of making energy, actually converts less than 1 percent of matter's mass-energy into light. In contrast, up to about 40 percent of the mass-energy can be converted into light when matter falls into a black hole. Given this efficiency, calculations show that the center of a young galaxy can shine as a quasar when its central, supermassive black hole consumes a mass equivalent to only about a dozen Suns each year.

Although it now seems clear that quasars shine through the energy released as matter falls into their supermassive black holes, many mysteries remain about quasars and the galaxies that contain them. For example, what effect does the energy output of the quasar have on the stars and gas in the surrounding galaxy? Why are quasars more common among young galaxies? When in a galaxy's history does its supermassive black hole form? And, perhaps most puzzling, what causes a quasar to shut down as a galaxy ages?

4

This photograph shows the inside of the Super-K neutrino detector in Japan before it was filled with water. The bright spots are some of the 11,000 phototubes in the detector. The total surface area of the phototubes is nearly an acre.

5

The 2.5-meter diameter telescope for the Sloan Digital Sky Survey (far left) is part of the Apache Point Observatory in the Sacramento Mountains of New Mexico. The white streak visible in the distance—just above the telescope in the photograph—is White Sands National Monument. (Upper right) This photograph shows astrophysicist Rich Kron inserting optical fibers into one of the pre-drilled plug plates (white disk). Each plate holds 640 optical-fiber cables that allow the telescope to obtain spectra of 640 galaxies simultaneously.

6

The Hubble Space Telescope extends outward from the Space Shuttle cargo bay during the 1997 servicing mission in which Shuttle astronauts replaced several instruments.

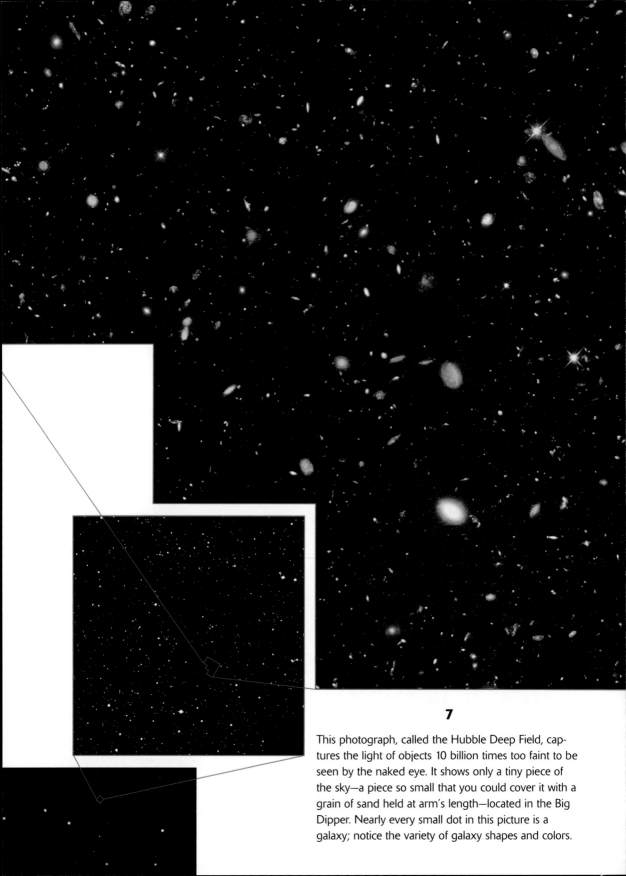

7

This photograph, called the Hubble Deep Field, captures the light of objects 10 billion times too faint to be
seen by the naked eye. It shows only a tiny piece of
the sky—a piece so small that you could cover it with a
grain of sand held at arm's length—located in the Big
Dipper. Nearly every small dot in this picture is a
galaxy; notice the variety of galaxy shapes and colors.

8

This artist's conception shows the power source of a quasar. A supermassive black hole resides in the center, surrounded by a disk of swirling material that will eventually fall into it. The jets emerging from either side of the disk are driven by the tremendous energy released as matter falls into the black hole. Everything shown in this picture would lie deep in the heart of a distant galaxy, and no telescopes are yet powerful enough to resolve details in such small and distant regions. (Painting by Joe Bergeron.)

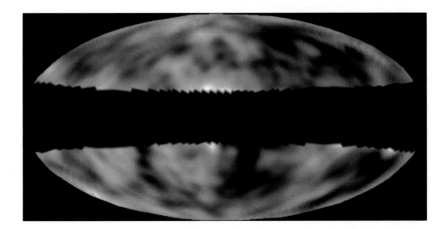

9

This image shows the "lumpiness" in the cosmic microwave background detected by the COBE satellite. The oval represents the entire celestial sphere, in much the same way that a flat map of Earth represents the entire globe. The black swath down the middle indicates the plane of the Milky Way galaxy, where data about lumpiness are currently unavailable. The regions above and below this swath represent directions looking north and south of the galactic plane. The image shows slight variations in the temperature of the cosmic microwave background, which correspond to differences in the density of the early universe.

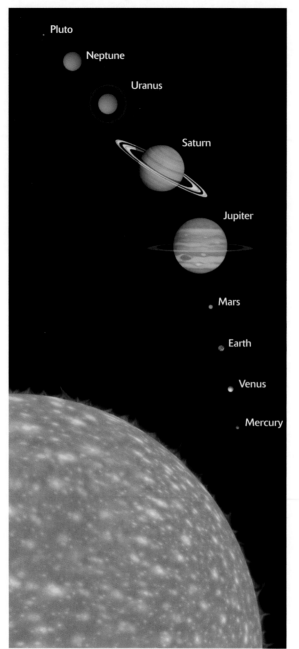

Pluto

Neptune

Uranus

Saturn

Jupiter

Mars

Earth

Venus

Mercury

10

This painting shows the Sun and the planets of our solar system at one ten-billionth of their actual size. (Distances are *not* to scale.)

11

The Very Large Array in Socorro, New Mexico, consists of 27 radio telescopes that work together as an interferometer. Their total light-collecting area is just the sum of the areas of the 27 individual dishes, but their combined angular resolution is equivalent to that of a single telescope that covers the entire area of the array.

12

The domes for the two 10-meter Keck telescopes on Mauna Kea, Hawaii. The two can work together as an interferometer, achieving much higher angular resolution than either telescope separately.

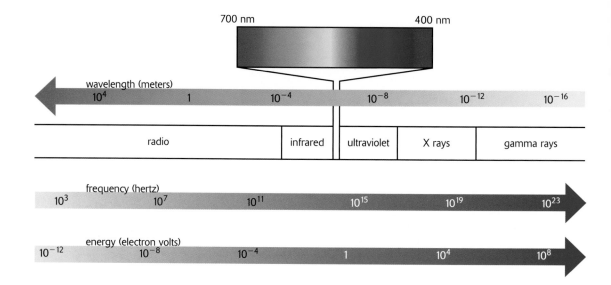

13

The complete electromagnetic spectrum ranges from low-energy radio waves to high-energy gamma rays. Note that the higher the energy, the higher the frequency and the shorter the wavelength. Our eyes can see only the small swath of visible light, but we can build detectors to record other forms of light.

That quasars do shut down is clear from the fact that they are much more common at larger lookback times, where we see galaxies as they were in the past, than nearby, where we see present-day galaxies. But the supermassive black holes themselves cannot simply disappear. If galaxies had such black holes in the past, the black holes must still be there today, and there do indeed seem to be supermassive black holes in the centers of many nearby galaxies. (Even our own Milky Way appears to harbor a very massive black hole in its center, though it is much less massive than those in quasars.) Though some of these galactic centers are still magnificently powerful, they do not shine as brightly as quasars, presumably because they are no longer sacrificing as much mass at the altars of their supermassive black holes.

No one knows what slows the consumption of matter by aging supermassive black holes, but one theory points to collisions. It may be that a supermassive black hole usually resides fairly quietly in its galaxy's center, consuming some of the surrounding matter at a slow but steady pace. A collision with another galaxy could supply it with a feast of extra gas, causing the center to brighten into a quasar. Once the feast was over, the supermassive black hole would return to its usual, more gradual pace of eating. Besides explaining how quasars turn on and off, this idea also explains why quasars were more common in the past: the fact that the universe is expanding tells us that galaxies must have been closer together in the past, so collisions between galaxies must have been more common.

Overall, using telescopes as time machines has shown us that galaxies go through remarkable changes during their lives. While we have ideas about the causes of many of these changes, such as what might change a spiral galaxy to an elliptical or what might cause a quasar to shut down, most of these ideas remain tentative. Coming observations with a new generation of powerful telescopes should shed much light on these issues. With luck, by the end of the first decade or two of the new millennium, we should be able to put together a clear and coherent story of how galaxies have evolved over the past 10 billion or more years.

The remaining pieces in the puzzle of galaxy evolution are the very earliest stages of galaxy formation and the seeds around which gravity acted to form galaxies. Again, forthcoming observations should answer many of the questions.

To learn about galaxy formation, we must find a way to observe very young—and hence very distant—galaxies and protogalactic clouds. We expect such objects to be difficult to see in visible light because they have few if any stars. However, they ought to show up with a powerful infrared telescope for two reasons. First, because very distant objects have such huge redshifts that any visible light they do emit is shifted into the infrared by the time it reaches Earth and, second, because very young stars tend to be enshrouded in gas that absorbs their visible light, leaving infrared light for us to detect.

NASA is planning to build such a telescope as the successor to the Hubble Space Telescope. Currently dubbed the *Next Generation Space Telescope,* or NGST, it could be built and launched as early as 2008. It will have a primary mirror at least twice as large as that on the Hubble Space Telescope, allowing it to see fainter objects. And because it will be optimized for infrared observations, it will be especially well suited to the task of seeing very young galaxies and protogalactic clouds in the far reaches of the observable universe.

In the meantime, astronomers have a less direct way of observing at least some protogalactic clouds. In much the same way that we can see dust suspended in the air when it is illuminated by a flashlight beam, we should be able to "see" protogalactic clouds if they are illuminated from behind. The ideal sources of illumination are distant quasars. As their bright light passes through clouds of gas that lie between us and them, some of the light is absorbed, leaving its mark in the quasar spectra. We need only to read the spectra carefully to learn about the clouds that lie in the light's path.

The major difficulty with this technique is that we expect most protogalactic clouds to have existed only when the universe was very young, since at later dates they would already have collapsed into full-fledged galaxies. We therefore can hope to see protogalactic clouds only in the spectra of unusually distant quasars—those that predated at least some protogalactic clouds in the early universe. So far, only a few candidates for protogalactic clouds have been found in such quasar spectra. But as we discover increasing numbers of distant quasars, we should also be able to find more and more protogalactic clouds. Such discoveries are now coming rapidly. During its first six months of operation alone, the Sloan Digital Sky Survey (see Mystery 8) discovered four quasars more distant than any known previously. Observations with NASA's Chandra observatory, an X-ray telescope that was launched into a high Earth orbit from the Space Shuttle in 1999, may also help astronomers discover distant quasars, since the gas falling into their supermassive black holes is so hot that it emits huge amounts of X rays.

Thus, assuming that no major surprises pop up as we fill in the details of galactic evolution and learn about the early stages of galaxies as protogalactic clouds, we can hope that the mystery of the existence of galaxies will soon be resolved to the point of the seeds. For this final question—the nature and origin of the seeds themselves—we must look as far back as possible into the history of the universe.

At first, you might think that we would be at a loss to see anything beyond the most distant quasars. After all, we would be looking back to a time before there were any bright objects at all in the universe. However, as we will discuss in more detail in Mystery 4, the Big Bang itself should have produced a lot of light. As it turns out, we can indeed detect this light from the creation of the universe. We "see" it with radio telescopes, and we call it the *cosmic microwave background.*

The cosmic microwave background appears to fill the entire universe, because it comes from all directions. No matter what direction we look, if we look far enough we should see to a lookback time shortly after the Big Bang. We cannot see all the way back to the beginning, because light was unable to travel freely until the universe was about 300,000 years old: prior to that the universe was so dense that light was absorbed by nearby matter as quickly as it was emitted. Thus the cosmic microwave background gives us a snapshot of what the universe must have looked like a mere 300,000 years after the Big Bang. This light was in the visible portion of the spectrum at the time it was released, but in accord with Hubble's law it has since been redshifted into the microwave portion of the spectrum (microwaves are relatively short wavelength radio waves).

If the seeds of galaxies were really present in the early universe, they should have left their marks on the cosmic microwave background. If instead the universe had been perfectly smooth when it began, then the cosmic microwave background would also be perfectly smooth. In the early 1990s, data from a NASA satellite called the Cosmic Background Explorer, or COBE (pronounced "ko-bee"), proved that the cosmic microwave background is *not* perfectly smooth (see Color Plate 9). However, the resolution of the COBE data is fairly crude, and the sizes of the lumps seen in its data are too large to be the seeds of individual galaxies. To confirm that seeds of the right size were present, we need higher-resolution pictures of the cosmic microwave background. Several planned experiments with ground-based radio telescopes and balloon-borne observatories should provide some such pictures, but two planned space missions probably hold the key to getting the necessary data.

If all goes well, by mid-2001 NASA will be operating a new satellite called the Microwave Anisotropy Probe, or MAP. Among its capabilities, MAP will be able to see the cosmic microwave background with roughly fifty times the detail captured by COBE. A few years later, a mission called Planck should be able to make a similar improvement on the resolution of MAP. Planck is being developed by the European Space Agency, which hopes to launch it by 2007. Although neither MAP nor Planck will be able to resolve structures as small as the seeds of galaxy-size objects, their resolution should be sufficient to allow astronomers to test competing models of the early universe. In that way, between MAP and Planck we ought to be able to determine whether the seeds of galaxies really were planted in the Big Bang.

Of course, even if the new missions confirm the existence of the seeds, we will still be left with the question of where the seeds themselves came from. Because we cannot see the events of the Big Bang, we will have to study this question with theoretical tools, which means this question must wait until we explore the theory of the Big Bang in Mystery 4. As we will see, this final piece in the mystery of the evolution of galaxies is tied up with the mystery of how the universe as a whole came to be.

*Do there exist many worlds,
or is there but a single world? This is
one of the most noble and exalted
questions in the study of Nature.*

ST. ALBERTUS MAGNUS (1206–1280)

Are Earth-like Planets Common?

The first definitive evidence of planets in other star systems came only in 1995, but today the known "extrasolar planets" already outnumber the planets in our own solar system. Announcements of such discoveries are now commonplace, sometimes giving the impression that we know all about other solar systems. But so far all of these new planets have proved much too large to be Earth-like. Thus we still don't know whether Earth-like planets are common in other star systems—or, indeed, whether they exist at all. In this mystery, we will discuss why scientists believe that Earth-like planets should be common and how missions planned for the first two decades of the third millennium will enable us to hunt for these planets. We'll also discuss a bit of the history of human understanding of planets and some of the technologies currently being used to detect planets in other star systems.

In popular films such as those in the *Star Trek* and *Star Wars* series, the galaxy is filled with Earth-like planets upon which humanoid civilizations make their homes. But are such planets really common? Although you've probably heard some of the many recent announcements of planets discovered around other stars, none of the planets discovered to date qualify as Earth-like. In other words, we could not live on them. The question of whether Earth-like planets are common remains open, qualifying as our Mystery 6.

Until a few hundred years ago, the idea of an Earth-like planet was an oxymoron. The Earth and the planets were thought to be creatures of totally different types. The Earth was the center of the universe, whereas the planets were mysterious bright stars that wandered among the other stars in the sky, which were fixed in the patterns of the constellations. Indeed, the word *planet* comes from the Greek for "wanderer." The planets held a special place in the mythology of almost every ancient culture. Often they were associated with gods, and special alignments of the planets were almost universally seen as some kind of omen.

Ancient people counted five planets wandering in seemingly unpredictable ways among the stars: Mercury, Venus, Mars, Jupiter, and Saturn. Each of these five planets is easily visible to the naked eye, though Mercury can be seen only rarely since it lies so close to the Sun. If you want to see Mercury, check a sky calendar to find out when it is positioned in a way that allows it to be seen shortly before sunrise or after sunset. (Sky calendars are published monthly in magazines such as *Sky and Telescope, Astronomy,* and *Mercury,* and they are also available on the Web.) The ancients did not know about Uranus, even though it is faintly visible to the naked eye, because no one noticed that it, too, wanders slowly among the stars. Nor did they know about Neptune or Pluto, because these worlds cannot be seen at all by the naked eye.

Because the Sun and Moon also obviously moved among the stars, though in much more predictable ways, they were counted as "planets" by the ancient Egyptians, Greeks, Romans, and other civilizations of the region. Thus these cultures recognized a total of seven planets, which is why we have seven days in a week. You can still see how these planets were honored if you use a bit of linguistic history. The honorees for Sunday, Monday, and Saturday are fairly obvious. For the other days, the correspondence in English is clear if you look at the old Teutonic names for the planets: Mars was Tiw (Tuesday), Mercury was Woden (Wednesday), Jupiter was Thor (Thursday), and Venus was Fria (Friday).

The planets were mysterious because they move among the stars in what seems a bizarre way—at least if you think that the Earth is the center of the universe. As we discussed in the beginning of Mystery 8, the Earth appears to be surrounded by a celestial sphere dotted with stars. This appearance makes it easy to explain the daily rising and setting of the stars by simply assuming that the celestial sphere turns daily around the Earth.

The motions of the Sun and Moon are only slightly more difficult to explain. Each day, the Sun moves slightly eastward relative to the stars, which is why the constellations rise a few minutes earlier each night. Over the course of a year, the Sun makes its way all the way around the celestial sphere, passing through the familiar constellations of the zodiac. The ancient Greeks explained this motion by imagining that the Sun resided on its own celestial sphere. By assuming that the Sun's sphere rotated daily around the Earth at a slightly slower rate than the sphere of the stars, they could explain why the Sun appeared to move gradually eastward relative to the stars.

The Moon also moves steadily eastward among the stars, taking roughly a month (think "moonth") to move all the way around the celestial sphere and complete a cycle of phases, from new moon to full moon and back again. Because this means that the Moon moves about one-thirtieth of the way around the celestial sphere each day, you can notice the Moon's eastward drift relative to the stars even over a few hours on a single night; just look closely to see how the Moon's position in the sky changes relative to some bright star. For the ancient Greeks, explaining and predicting the Moon's motion required nothing more than adding a third celestial sphere to carry the Moon through the heavens.

The motion of the planets is not explained so easily. If you observe carefully, you can see the planets drifting relative to the stars from one night to the next—or from one week or month to the next, in the case of the more distant planets. However, unlike the Sun and Moon, the planets do not always move eastward relative to the stars. Sometimes they turn around and head westward in what we call their *apparent retrograde motion*. Because each planet moves differently, the Greeks added another celestial sphere for each one, but this ploy was unsatisfying because they had to imagine that the planetary spheres kept

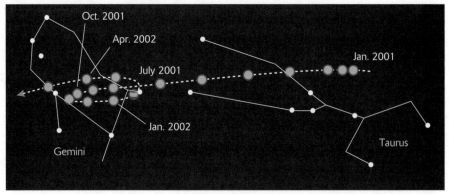

Dots represent Jupiter's position at one-month intervals.

Jupiter's position among the stars in our sky in 2001–2002. Notice that Jupiter generally drifts eastward relative to the stars, but for a few months each year it goes westward in its apparent retrograde motion. Jupiter appears to move through roughly one constellation of the zodiac per year, because it takes twelve years to orbit the Sun.

reversing their directions of rotation. Moreover, because they could not explain when or why these spheres changed direction, the added spheres were useless for predicting planetary positions.

The Greeks struggled to explain planetary motion for hundreds of years. For a while—well, for about fifteen hundred years—they appeared to succeed. In a triumph of human ingenuity, the famed Greek astronomer Ptolemy found a way to predict planetary motion while preserving the idea of an Earth-centered universe. Working around AD 150, he synthesized and extended earlier Greek ideas to create a model of the universe that could be used to predict planetary positions in the sky within a few degrees, which was good enough to agree with the naked-eye observations of the time.

Although the Ptolemaic model worked well, it was extremely complex. Following a Greek tradition dating back to Plato (428–348 BC), Ptolemy imagined that all heavenly motions must be in perfect circles. To explain the apparent retrograde motion of planets, Ptolemy used an idea first suggested by Apollonius (c. 240–190 BC) and further developed by Hipparchus (c. 190–120 BC). This idea imagines that planets move along small circles that, in turn, move in a larger circle around the Earth. As seen from Earth, this "circle upon circle" motion caused

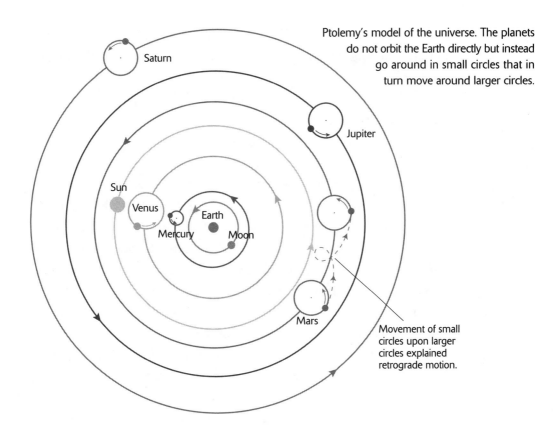

Ptolemy's model of the universe. The planets do not orbit the Earth directly but instead go around in small circles that in turn move around larger circles.

Saturn

Jupiter

Sun

Venus

Earth

Mercury

Moon

Mars

Movement of small circles upon larger circles explained retrograde motion.

a planet to move sometimes eastward and sometimes westward relative to the stars.

By adjusting the sizes of the circles and the rates of motion of the planets along them, Ptolemy's model could reproduce the apparent retrograde motion of all the planets. To fine-tune his predictions, Ptolemy also incorporated a few more complexities into his model, such as allowing the Earth to be slightly off-center from the circles of a particular planet. A thousand years later, while supervising calculations based on the Ptolemaic model, the Spanish monarch Alfonso X is said to have remarked, "If I had been present at the creation, I would have recommended a simpler design."

Ptolemy's model was not only complex, but by the late 1400s its predictions were failing noticeably. This failure provided the backdrop for a rethinking of planetary motion, and here is where Nicolaus Copernicus made his mark on history. Born into a wealthy Polish family in

1473, Copernicus turned his attention to astronomy in his late twenties after having first studied painting, mathematics, medicine, and law. Drawing on the idea suggested nearly eighteen hundred years earlier by Aristarchus, Copernicus concluded that apparent retrograde motion was much more easily explained by removing the Earth from its exalted place in the center of the universe.

Once we accept the Copernican idea that the Earth is a planet going around the Sun, we can understand apparent planetary motion without requiring anything ever to go backward. You can see how with the help of a friend. Go outside, to a lawn or a parking lot or any other open area, and pick a spot in the center to represent the Sun. You'll represent Earth by walking around the Sun, while your friend represents Mars, Jupiter, or Saturn by walking around the Sun more slowly and at a greater distance. If you watch closely as you walk, you'll see that your friend appears to move backward compared to background objects during the part of your "orbit" in which you catch up and pass him or her. In the same way, the planets do not really go backward in their apparent retrograde motion but only appear to do so as the Earth passes them in its orbit. (You can see how apparent retrograde motion works for Mercury or Venus by switching places with your friend.)

Despite the simplicity of his explanation for planetary motion, Copernicus did not win the day immediately. One reason was that he still had not explained the absence of stellar parallax, as discussed in Mystery 8. But the bigger reason was that his model didn't make much better predictions of planetary motion than the Ptolemaic model. The problem was that Copernicus still used the Platonic idea of perfect circles for his planetary orbits. Because orbits are in fact elliptical, his model made poor predictions. Nevertheless, Copernicus started a scientific train that could not be stopped, and within a hundred years the work of Tycho Brahe, Johannes Kepler, and Galileo finally laid the Earth-centered model to rest. By the end of the seventeenth century, Isaac Newton had discovered the law of gravity, and with his new mathematics of calculus, it became possible to predict planetary motion with great precision.

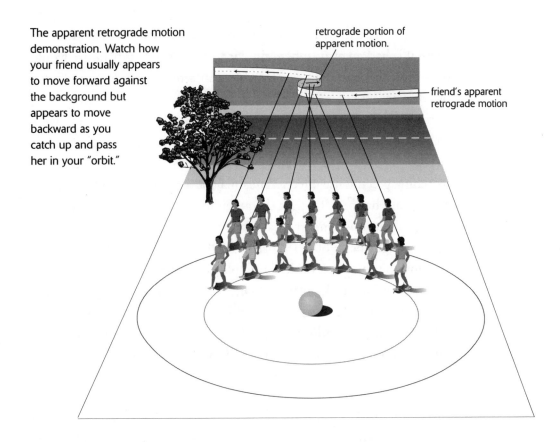

The apparent retrograde motion demonstration. Watch how your friend usually appears to move forward against the background but appears to move backward as you catch up and pass her in your "orbit."

retrograde portion of apparent motion.

friend's apparent retrograde motion

Placing the Earth in its proper class as a planet almost immediately raised the question of whether other Earth-like planets existed. While people had speculated about life on the Sun, Moon, and planets before—some cultures have stories that go back thousands of years—these speculations always had a mythological quality. The Copernican revolution put the question of Earth-like planets firmly in the realm of science.

For a while, nearly all the planets were considered potentially Earth-like. It was not until the 1800s, when astronomers finally worked out the true scale of the solar system, that it became clear that only Venus and Mars were at all Earth-like in size. Venus might have seemed the better candidate to be Earth-like in character, because it is almost identical in size to Earth and not that much closer to the Sun. Indeed, many scientists and writers imagined Venus as a tropical paradise, because if it had an atmosphere similar to Earth's its temperature would be uniformly warm and comfortable. But attention ended up focused on Mars because of its supposed canals (see Mystery 10). By the 1960s, hopes of finding an Earth-like planet in our own solar system were dashed. Spacecraft had conclusively proved that the Martian canals were an illusion, and spectroscopic studies of Venus had revealed its scorching greenhouse effect. Venus may be a great place to learn about what could happen if we let the greenhouse effect get out of control on Earth, but it will never be a popular vacation spot.

Through the 1960s, 1970s, and 1980s, astronomical technology was not up to the task of detecting planets around other stars, and the question of Earth-like planets—or, for that matter, of any planets at all beyond our solar system—was primarily theoretical. This theoretical front saw tremendous progress, thanks in large part to data from the Apollo missions to the Moon and the first generation of planetary probes. Today we believe we have a very good theory of how our solar system formed from a cloud of interstellar gas and dust in the Milky Way. This theory not only explains many of the characteristics of our solar system but also predicts that such systems ought to be fairly common.

In brief, the theory of solar system formation holds that all stars form in collapsing clouds of interstellar gas and dust. The composition of the clouds reflects the overall chemical composition of the Milky Way galaxy, which is roughly 98 percent hydrogen and helium, and 2 percent heavier elements. As gravity collapses a cloud, it spins faster and flattens into a disk—just as protogalactic clouds do (see Mystery 7). A star forms in the center of the disk, where the temperature and density are highest, while the gaseous disk continues to rotate around it. As temperatures in the disk cool, solid particles can condense from the gas and collide with each other, sometimes sticking together. As these conglomerates (called *planetesimals*) grow bigger, their gravity attracts more and more material. Over a few tens of millions of years, they grow into full-fledged planets. The star's "wind," which consists of particles blown off its surface, sweeps any remaining gas into interstellar space, leaving behind a solar system with a star and planets.

This theory makes it easy to explain why the inner planets of our solar system (Mercury, Venus, Earth, and Mars) are so different in character from the giant outer planets (Jupiter, Saturn, Uranus, and Neptune). Remember that 98 percent of the gaseous disk out of which planets will form is hydrogen and helium. Much of the other 2 percent is carbon, nitrogen, and oxygen, and these elements can combine with hydrogen to make fair quantities of compounds like methane (CH_3), ammonia (NH_3), and water (H_2O). This leaves only a very small fraction of the gas cloud in the form of elements that make up rock and metal (such as silicon, nickel, and iron).

In the inner regions of a gaseous disk, near the central star, it is too hot for the water, ammonia, or methane to condense into ice. But because metal and rock can remain solid at fairly high temperatures, bits of solid metal and rock condense from the warm gas. Because the metals and rock are so rare, there's not enough solid material in these inner regions to make planets much larger than the Earth. Moreover, these small planets lack the gravitational strength needed to hold onto the abundant gas around them.

Bits of metal and rock also condense in the outer regions of the gaseous disk, but here the temperatures are so cold that the more abundant compounds of water, ammonia, and methane also condense—into pieces of solid ice. Thus the growing chunks of solid material in an outer solar system are made mostly of ice mixed with smaller amounts of metal and rock. These iceballs can grow fairly large—perhaps ten times the size of the Earth or more. At that point, their gravity begins to attract the surrounding hydrogen and helium gas, which makes them grow even bigger. By the time the process is complete, these outer planets have become giants made mostly of hydrogen and helium. The gas drawn into a large planet tends to form a spinning disk, rather like a miniature version of the spinning disk around the star. The same general processes then occur within these planetary disks, leading to the formation of moons around these planets just as the planets formed around the star—one reason why the giant outer planets of our solar system tend to have many moons. (Pluto is in a different category than any of the other planets. It is the smallest planet by far and made mostly of ice, and today it is thought to be just the largest of thousands of similar, comet-like bodies in the outer reaches of the solar system.)

You can probably imagine several ways in which a wrench could be thrown into this planetary formation process. For example, if the central star rotates too quickly as it is forming, it will split in two, becoming a binary star system. The competing gravitational attraction of the two companion stars might disrupt the disk and prevent planets from forming. This type of splitting apparently occurs about half the time, since roughly half the star systems in our galaxy consist of two (or more) stars. However, some recent evidence, both observational and theoretical, suggests that planets can form even in such cases. Another problem might arise if a young star has a particularly strong stellar wind, in which case it might blow away the gas in its disk before planets have a chance to form. A third problem involves time: it takes tens of millions of years to form planets, and some stars don't live that long. Nevertheless, given enough time and a single star like our Sun, our theory of solar system formation makes the creation of planets seem almost inevitable. Because most stars are long-lived (billions of years), planets ought to be extremely common.

Unfortunately, the task of finding planets around other stars is dauntingly difficult. You can see why by visualizing a scale model of our solar system (see the photograph on the facing page and Color Plate 10). Imagine the Sun as a grapefruit (which makes it about one ten-billionth of its actual size). On this scale, Earth is the size of a pinhead and orbits the Sun at a distance of about 15 meters (16.5 yards). The largest planet, Jupiter, is about the size of a marble and orbits the Sun at a distance of a bit less than the length of a football field. Pluto is smaller than the period at the end of this sentence and about 600 meters (650 yards) from the model Sun. If you remember that the planets shine only by reflected sunlight, you'll understand why Pluto cannot be seen without a telescope, even though it is within our own solar system. But this is nothing compared to the challenge of seeing planets in another star system, because, on the same scale, the stars nearest the Sun—the three stars of the Alpha Centauri system—are more than 4,000 kilometers (2,500 miles) away!

Trying to see an Earth-size planet in the Alpha Centauri system is like looking out your window in New York and trying to see a pinhead in Los Angeles. All other stars are even farther away, so it's hardly surprising that we have not yet detected any Earth-like planets around other stars. Indeed, the bigger surprise may be that we *have* discovered Jupiter-size planets (and even Saturn-size planets) and that we should have the technology to find Earth-like planets within a couple of decades. It is time to turn our attention to the remarkable technologies that make planet hunting possible.

The first clear-cut discovery of a planet around another Sun-like

The scale-model solar system on the campus of the University of Colorado, Boulder, shows sizes and distances in our solar system at one ten-billionth of their actual values. Here we see the grapefruit-size Sun atop the pyramid. The model planets are housed in smaller granite pedestals along a walkway extending to the left. The child is the author's son, getting his first lesson about the scale of the universe.

star—a star called 51 Pegasi—came in 1995. The discovery was made by Swiss astronomers Michel Mayor and Didier Queloz, and it was soon confirmed by a team led by Geoffrey W. Marcy and R. Paul Butler of San Francisco State University. Many more extrasolar planets have been discovered since that time, most of them by the team led by Marcy and Butler. If you've followed the news about these planets, you may have been overwhelmed by the details associated with many of the planet-hunting strategies. However, if we strip away the details, there are really only two basic ways to search for extrasolar planets: directly and indirectly. Direct searches attempt to take pictures of the planets themselves. Indirect searches look only at stars, searching for effects that can be attributed to orbiting planets.

All extrasolar planet discoveries to date have been indirect. The most common indirect method looks for evidence of the gravitational tug that a planet exerts on its star. Although we usually think of a star as remaining still while planets orbit it, the truth is that all objects in a star system, including the star itself, orbit the system's center of mass. In the case of our own solar system, the Sun is so much more massive than any planet that the system's center of mass lies well inside the Sun—but not exactly at the Sun's center. In principle, then, the Sun's center moves around this center of mass in a complex way determined by the combined effects of all the planets. However, because Jupiter is so much more massive than any of the other planets (it outweighs the rest of the planets combined), nearly all of this motion is due to Jupiter. Thus the Sun's center traces a small circle around the center of mass with the same twelve-year period that Jupiter orbits the Sun. An observer in another star system could therefore infer the existence of Jupiter by noticing how the center of the Sun "wobbles" about the center of mass.

In a similar way, we can look for planets in other star systems by carefully watching for periodic wobbles in a star's position in the sky. One way to look for this wobble is to make very precise measurements of a star's position (astronomers refer to this technique as "astrometry"). But it's usually easier to observe Doppler shifts in a star's spectrum. As we discussed in Mystery 8, a Doppler shift arises when an object is moving relative to us. The light of an object moving toward us is shifted to shorter (bluer) wavelengths, while the light of an object moving away from us is shifted to longer (redder) wavelengths. Thus, if a planet is causing a star to wobble around a center of mass, the starlight should show a slight blueshift when the star is coming toward us and a slight redshift when the star is moving away from us. The main

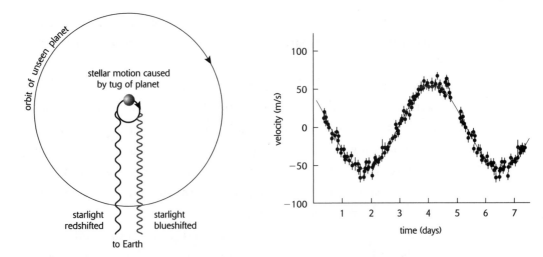

(Left) Doppler shifts allow us to detect the slight wobble of a star caused by an orbiting planet. (Right) This figure shows many measurements of the speed of the star 51 Pegasi inferred from Doppler shifts in its spectrum. (Dots are data points, bars are uncertainty in the measurements, and the curve is a theoretical fit to the data.) The star's speed clearly follows a pattern indicating a small orbital motion with a period of about five days. We can conclude that this wobble is caused by an orbiting planet with a mass at least 40 percent that of Jupiter and an orbital period of five days.

difficulty is that the shifts are very slight: typically, the tug of a planet will cause a star to wobble around the center of mass at only about 60 miles per hour, which is extremely slow by astronomical standards. Nevertheless, modern technology enables astronomers to measure the Doppler shifts even from such low speeds.

The time it takes for each cycle of blue- and redshifts tells us the orbital period of the planet that is causing the motion, and the precise pattern of the Doppler shifts tells us how much the planet's orbit deviates from a perfect circle (the orbital eccentricity). We can also estimate the planet's mass, because a larger Doppler shift generally means that the star is pulled at greater speed by the planet, which means the planet must be more massive for a given orbital distance. However, there is one important caveat to these mass estimates: because the Doppler shift tells us only about the star's motion along our line of sight, we don't automatically know how fast the star is moving perpendicular to our line of sight. Thus the speed we measure with the Doppler shift is a lower limit to the star's actual speed in its small orbit, and hence the planetary mass we estimate from this technique is also a lower limit to the planet's actual mass.

If a star happens to have more than one planet exerting a significant gravitational tug, the star will wobble in more complex ways around the center of mass. Thus, by carefully analyzing a star's wobble, we can sometimes infer the existence of more than one planet. Such analysis was used in mid-1999 to infer the existence of three planets around the star Upsilon Andromeda, making this the first bona fide, multiple-planet solar system known beyond our own.

For the sake of completeness, we should mention three other indirect planet-hunting techniques you may have read about in news reports. One involves looking for transits in which a star dims slightly as a planet passes in front of it. The first such transit was detected in late 1999 for a planet orbiting the star HD 209458. A second technique involves looking at a star's spectrum for light absorbed or reflected by an orbiting planet. As the planet moves around the star, this light will have a slightly different Doppler shift than the light of the star itself, so a planet's existence can be revealed by the presence of such light in a star's spectrum. The third method involves an effect called microlensing, which occurs when a planet around a moderately distant star happens to pass directly in front of an even more distant star. In accord with Einstein's general theory of relativity, this can cause a temporary brightening of the distant star's light.

The chart on the facing page shows the planetary discoveries made as of early 2000. Note that all these planets are substantial in mass—Jupiter-size or larger—and most orbit quite close to their star. In light of what we've already said about solar system formation, this should seem odd; the inner part of a solar system ought to have only relatively small planets made of metal and rock. While this has made some astronomers wonder if there's something wrong with our theory of solar system formation, most think these odd discoveries can be explained without having to go back to the drawing board. The indirect techniques are biased toward the discovery of large planets in close orbits—

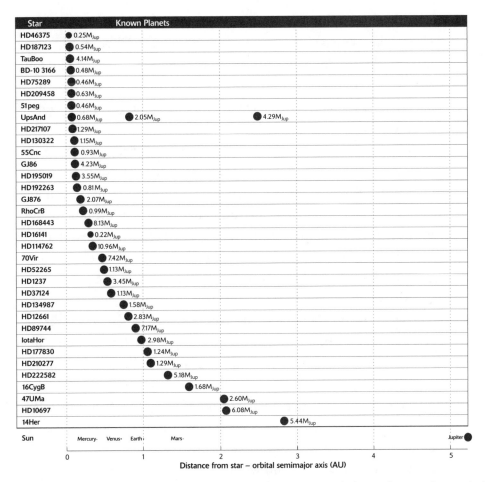

Star	Known Planets			
HD46375	● 0.25M$_{Jup}$			
HD187123	● 0.54M$_{Jup}$			
TauBoo	● 4.14M$_{Jup}$			
BD-10 3166	● 0.48M$_{Jup}$			
HD75289	● 0.46M$_{Jup}$			
HD209458	● 0.63M$_{Jup}$			
51peg	● 0.46M$_{Jup}$			
UpsAnd	● 0.68M$_{Jup}$	● 2.05M$_{Jup}$	● 4.29M$_{Jup}$	
HD217107	● 1.29M$_{Jup}$			
HD130322	● 1.15M$_{Jup}$			
55Cnc	● 0.93M$_{Jup}$			
GJ86	● 4.23M$_{Jup}$			
HD195019	● 3.55M$_{Jup}$			
HD192263	● 0.81M$_{Jup}$			
GJ876	● 2.07M$_{Jup}$			
RhoCrB	● 0.99M$_{Jup}$			
HD168443	● 8.13M$_{Jup}$			
HD16141	● 0.22M$_{Jup}$			
HD114762	● 10.96M$_{Jup}$			
70Vir	● 7.42M$_{Jup}$			
HD52265	● 1.13M$_{Jup}$			
HD1237	● 3.45M$_{Jup}$			
HD37124	● 1.13M$_{Jup}$			
HD134987	● 1.58M$_{Jup}$			
HD12661	● 2.83M$_{Jup}$			
HD89744	● 7.17M$_{Jup}$			
IotaHor	● 2.98M$_{Jup}$			
HD177830	● 1.24M$_{Jup}$			
HD210277	● 1.29M$_{Jup}$			
HD222582	● 5.18M$_{Jup}$			
16CygB	● 1.68M$_{Jup}$			
47UMa		● 2.60M$_{Jup}$		
HD10697		● 6.08M$_{Jup}$		
14Her		● 5.44M$_{Jup}$		
Sun	Mercury· Venus· Earth· Mars·			Jupiter ●

```
0          1          2          3          4          5
        Distance from star – orbital semimajor axis (AU)
```

This diagram shows the orbital distances and approximate masses of planets discovered around other Sun-like stars as of early 2000. The masses are lower limits to the actual masses of the planets.

mainly because such planets have greater gravitational effects on their stars than do smaller or more distant planets. Moreover, while it takes many years of observations to discover the wobble caused by a planet with a twelve-year period like Jupiter, the wobble caused by a close planet with an orbital period of days or months can be discovered in a correspondingly short time. Solar systems with large, close-orbiting planets may therefore be relatively rare, even though they are dominant among the ones we've discovered so far.

Still, even if they are rare, it would be nice to understand how such systems can form, and our theory of solar system formation allows several possibilities. One is that these large planets might not be "true" planets, in the sense of having formed in a gaseous disk around a star. Instead, they might have formed much as binary stars form, by splitting off from a fast-rotating central star. In that case, they are more like failed stars—stars too small to sustain nuclear fusion in their cores—than successful planets. Another possibility is that some of these planets formed around stars that had unusually weak stellar winds. In that case, the gaseous disk might have survived tens of millions of years longer than it did in our solar system, allowing it time to cool enough so that ices could condense and form large planets in the inner part of the star system. A third possibility is that some of these planets did form in their outer star systems but somehow migrated inward, perhaps as the result of a near collision with another large planet.

Regardless of how they came to exist, all the planets discovered to date seem far too big to be Earth-like in any way; their large sizes suggest that they are probably made mostly of gas, like Jupiter, rather than of rock. And while it is conceivable that some could have moons that are Earth-like, we have no way to know at present. Thus, while these planetary discoveries are exciting because they confirm that planets exist around other stars, they have not yet told us much about the question of Earth-like planets.

Resolving that question will require new observation strategies. One strategy involves looking for transits of planets passing in front of their stars. A proposed NASA mission called Kepler, which could be launched as early as 2005 if it is funded, would search for transits around some one hundred thousand stars. Kepler would be capable of detecting Earth-size planets (as well as larger planets); however, because transits can be seen only for planets that happen to orbit their star in the plane of our line of sight, this mission could detect only a small fraction (about one-half of one percent) of the planets among its target stars. A more general search strategy involves building a telescope that can detect very tiny wobbles in a star's position—much smaller than the wobbles caused by Jupiter-size planets—thereby enabling us to conclude that the star is being tugged by an object as small as the Earth.

To see such tiny wobbles, we'd need a telescope with very high *angular resolution,* or ability to see small details in images. The angular resolution of a telescope depends on a number of factors, including the quality of its optics and the stillness of the atmosphere around it (not a factor for telescopes in space), but it is ultimately limited by the telescope's size: all other things being equal, larger telescopes have smaller and hence better angular resolution. Unfortunately, the angular resolution required to detect the wobbles caused by Earth-size planets is so small that, with current cost constraints, we have little hope of building a big enough telescope (at least in space). However, a remarkable technique called *interferometry* may soon allow us to achieve the necessary resolution through simultaneous observations made by two or more telescopes linked together.

In essence, interferometry allows two or more telescopes to achieve the angular resolution of a much larger telescope. The technique takes advantage of the wavelike properties of light by observing how light waves collected by the individual telescopes interfere with one another when combined. Although the details of interferometry are fairly complex, the idea is somewhat like watching the ripples on a pond reaching the shore in two different places. In principle, the timing of the ripples' arrival tells you something about where they came from. Interferometry is much easier to implement with longer-wavelength light and was first achieved with radio telescopes, since radio waves have the longest wavelengths of all forms of light (see Color Plate 11). Recently, however, astronomers have had some success in extending this technique to infrared wavelengths and are working on implementing interferometry for visible wavelengths as well. Several experiments with ground-based interferometers are already under way, including one that uses the twin 10-meter Keck Telescopes on the summit of Mauna Kea, in Hawaii (see Color Plate 12). Some of these experiments may soon be able to detect planets as small as Uranus or Neptune, but not as small as Earth, around nearby stars.

Part of the challenge of interferometry lies in the precision timing of light waves that it requires. For infrared and visible wavelengths, even the slightest atmospheric turbulence can make this precision impossible to obtain. As a result, astronomers hope to place interferometers in space, where they will be immune to the Earth's tremors as well as to the blurring effects of the Earth's atmosphere.

The Space Interferometry Mission, or SIM, is the first of two key optical interferometry missions planned by NASA. Depending on its final configuration, SIM may be capable of detecting the small stellar wobbles caused by planets of Earth's size or slightly larger around the nearest few dozen stars, and somewhat larger planets around somewhat more distant stars. If all goes well with its technology development and funding, SIM may be launched as early as 2006; thus, between SIM and the Kepler mission, the first evidence of Earth-size planets could come in this decade. However, even if Kepler or SIM detect a few Earth-*size* planets, we still won't necessarily know whether they are Earth-like. For that, we need to study the planets directly.

The planned follow-up mission to SIM, called the Terrestrial Planet Finder, or TPF, may finally resolve the mystery of whether Earth-like planets are common. As currently envisioned, TPF will consist of four 3.5-meter telescopes working together as an interferometer to achieve an angular resolution equivalent to that of a single telescope with a diameter of more than 100 meters. In principle, the four telescopes together will have sufficient light-collecting area to see the dim infrared light emitted by Earth-size planets around nearby stars, and the angular resolution of the four-telescope combination will be sufficient to distinguish these planets from the stars themselves. TPF will also employ a new technology called *nulling* that will help it to blot out the bright light of the central star, which would otherwise drown out the dim light of an orbiting planet. (Nulling will also be used on SIM, but SIM is not expected to be able to image Earth-like planets.)

If all goes well, TPF will have the power to make a conclusive search for Earth-size planets around one hundred fifty nearby, Sun-like stars. If the planets are there, TPF should find them. Moreover, once an Earth-size planet is discovered, TPF will be able to record its spectrum, allowing us to determine whether the planet's composition is Earth-like. It should even be able to detect the presence of atmospheric gases that may be indicative of life (such as a combination of oxygen and methane). NASA hopes to launch TPF by about 2011. The European Space

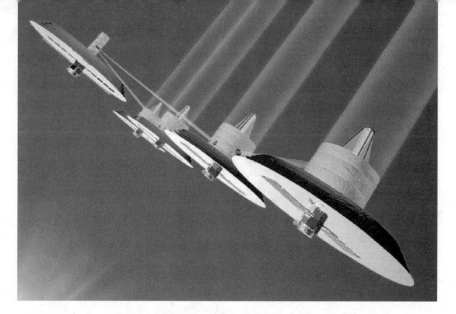

This rendering shows one possible configuration for the Terrestrial Planet Finder. It consists of four separate telescopes flying freely in space. The central spacecraft performs the interferometry by combining the light from all four telescopes.

Agency is considering a similar mission for infrared interferometry, called Darwin, on roughly the same time scale.

Later in the twenty-first century, astronomers hope for even more powerful interferometers either in space or on the surface of the Moon. It's well within reason to imagine that during this century we'll see optical interferometers with dozens of telescopes operating across a diameter of hundreds of kilometers. Such telescopes will be able not only to detect Earth-like planets around nearby stars but also to obtain fairly clear images of those planets. We'll see whether they have continents and oceans like the Earth and perhaps gather clear evidence for life.

As implied by the quotation from St. Albertus Magnus at the beginning of this chapter, our Mystery 6 may have deep philosophical implications. What would it mean to know that Earth-like planets are common—or, conversely, that they are exceedingly rare? How would it affect our view of ourselves and of our planet? One way or the other, we'll soon find out, because barring a major failure of our space program, this mystery will be solved by the third decade of the third millennium.

The most exciting phrase to hear in science, the one that heralds new discoveries, is not "Eureka!" (I found it!) but "That's funny . . ."

ISAAC ASIMOV (1920–1992)

What Makes Gamma-Ray Bursts?

If today is a typical day, observatories in Earth orbit will at some point detect a burst of gamma rays—the highest-energy form of light—coming from the distant reaches of the cosmos. The strength of these gamma-ray bursts, along with their distant origins, stands as testimony to their incredible power, yet to date we have little idea of their cause. In this mystery, we will explore the mysterious origins of gamma-ray bursts, a search that will take us into the exotic realm of such bizarre objects as neutron stars and black holes.

For the most part, the universe moves at a stately and ponderous pace, and major events unfold over thousands or millions of years. Only occasionally do we see more dramatic, explosive events. The most famous cosmic explosions are supernovae—stars that explode in their entirety. For a few weeks, these exploding stars can shine nearly as brightly as an entire galaxy, making them visible to our telescopes even when they occur in distant reaches of the universe.

Supernovae, which were once a major mystery of their own, seem to be fairly well understood today. But what if astronomers suddenly began to discover explosions that dwarfed the known supernovae? Imagine an explosion that, for a few seconds, could release as much energy as a hundred quadrillion (10^{17}) Suns. Imagine that astronomers discovered such an explosion occurring somewhere in the universe almost every day. Imagine that, despite decades of effort, astronomers still had little idea of what caused these fantastic explosions. Now stop imagining, because this is our Mystery 5—the mystery of the so-called *gamma-ray bursts*.

A gamma-ray burst is pretty much what its name implies—a sudden and short burst of gamma rays, typically lasting between a few seconds and a couple of minutes. Gamma rays are a high-energy form of light that cannot be seen by the human eye. You are probably aware that many forms of light are invisible to our eyes, such as infrared or ultraviolet light. In fact, visible light is only a small part of the complete spectrum of light, which is known as the electromagnetic spectrum (because light waves consist of oscillating electric and magnetic fields). Scientists classify the different forms of light according to how much energy they carry (see Color Plate 13). Radio waves are the least energetic, followed in order of increasing energy by infrared and visible light. Ultraviolet light is more energetic than visible light, and X rays pack an even larger punch—which is why X rays can penetrate soft tissue, allowing doctors and dentists to make images of our bones and teeth. Gamma rays are at the highest energy end of the electromagnetic spectrum.

Generally speaking, the types of light emitted by an object depend on its temperature and the energy of any physical processes occurring within it. Cool objects, such as planets and people, emit mostly infrared light. Objects with temperatures of a few thousand degrees, such as the Sun's surface or the tungsten filaments in incandescent light bulbs, emit visible light. Stars that are hotter than the Sun emit mostly ultraviolet light. X rays come only from very hot material, such as the Sun's corona (its hot upper atmosphere), or from very energetic processes such as nuclear detonations or radioactive decay. Gamma rays, with their extraordinary energy, can come only from energetic particle interactions or from exotic cosmic powerhouses. Thus detecting a burst of gamma rays tells us that we are witnessing some kind of highly energetic explosion.

The mystery of gamma-ray bursts actually began as a mystery about sinister nuclear explosions. In August 1963, following eight years of discussion, the United States, Britain, and the Soviet Union signed the Treaty Banning Nuclear Weapon Tests in the Atmosphere, in Outer Space, and Under Water (commonly known as the limited nuclear test ban treaty). Although the treaty had provisions for monitoring compliance, including on-site inspections, the United States feared that the Soviets might still find a way to conduct secret tests. The United States therefore launched a series of satellites to watch for clandestine nuclear explosions. These satellites, given the name Vela (from the Spanish *velar,* "to watch"), were primarily designed to detect X rays emitted by a nuclear blast. However, some American strategists worried that the Soviets might find a way to hide the initial flash of X rays by detonating their nuclear tests behind a thick shield. Some even worried that the Soviets would conduct tests on the far side of the Moon! To insure against this possibility, they added gamma-ray detectors to the Vela satellites. Because gamma rays are produced in the cloud of radioactive material blown out by a nuclear blast, the satellites could thereby detect the debris from a nuclear explosion even if the blast itself were hidden.

The early Vela satellites saw no gamma rays, but as newer satellites in the series grew more sensitive, they began to detect occasional gamma-ray bursts. The precise signatures of the bursts did not look like what military scientists expected from nuclear explosions, so the treaty monitors were not too worried. Nevertheless, it was difficult to be sure, because they couldn't tell where the gamma rays were coming from.

The difficulty arose because gamma rays cannot be detected with ordinary telescopes. The basic idea behind a telescope is to collect a lot of light and bring it to a focus. Once it's in focus, we can see an image of the light source and where the source is located in the sky. This strategy works for almost all wavelengths. Like visible light, most infrared and ultraviolet light can be collected with ordinary glass mirrors. Radio waves can be collected and focused in a similar way, but because of their very long wavelengths, radio wave "mirrors" consist of a dish-shaped metal mesh rather than smooth, polished glass. X rays are somewhat more difficult to focus because they have sufficient energy to go through any mirror they hit straight on, but they can be focused with funnel-like telescopes, in which the X rays bounce along mirrors at shallow angles until they come to a focus at the end of the funnel (generally, only two bounces are required). Gamma rays, however, will penetrate any material at any angle, which makes them almost impossible to focus. Gamma-ray detectors therefore consist of thick blocks that gamma rays cannot go all the way through. Such a detector can record a gamma-ray strike but gives little information about where the gamma ray came from. Thus the gamma rays detected by the Vela satellites meant that an explosion was occurring *somewhere,* but for a while the scientists could not be sure whether it was somewhere on Earth or in some distant region of space.

Fortunately for concerned members of the military, the Vela satellites had been launched in pairs, with each member of a pair located at opposite sides of the same Earth orbit. Because the satellites shared the same high orbital altitude, gamma rays from an explosion in the Earth's atmosphere would have reached both satellites in a pair at almost precisely the same moment. But if the explosions were coming from some point in space, the gamma rays might reach one satellite slightly before the other—with a gap of up to about one second, since the diameter of the satellite orbits was close to the distance that light travels in one second (300,000 kilometers or 186,000 miles). By 1973, the military scientists were convinced that the explosions were of cosmic origin. They published their findings in an unclassified astronomical journal, taking the mystery out of the realm of the military and into astronomy. (Interestingly, Soviet satellites had also discovered gamma-ray bursts, and Soviet scientists concluded that the bursts were extraterrestrial in origin—and published their results in Russian astronomical journals—before their American counterparts.)

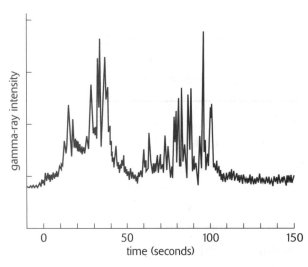

This graph shows the intensity of gamma rays recorded during a typical gamma-ray burst. Note that the intensity fluctuates dramatically during the burst, which lasts a total of about two minutes. Bursts similar to this one appear to occur about once a day on average and may, at their peaks, produce power equivalent to that produced by a hundred quadrillion Suns.

For a while, the gamma-ray bursts did not seem all that mysterious and would not have ranked among the top ten mysteries until almost two decades after their discovery. This might seem surprising, especially since most astronomers had previously assumed that no cosmic event was energetic enough to shine brightly in gamma rays. But around the time of the gamma-ray burst discovery, astronomers also discovered similar-looking bursts of X rays. The sources of these X-ray bursts were soon identified, and many astronomers assumed that the gamma-ray bursts came from similar sources.

The sources of X-ray bursts are no longer mysterious, but they stretch our imaginations nonetheless. These bursts come from binary star systems in which one member is a neutron star—an exotic type of "dead" star that is made almost entirely from neutrons. Neutron stars are bizarre in many ways, but for our discussion here their important property is that they are extremely dense—a teaspoon of matter from a neutron star would outweigh Mount Everest. A typical neutron star is no larger in diameter than a small city, but it contains more mass than our entire Sun. This compactness makes gravity incredibly strong near the surface of a neutron star. If a neutron star happens to be in a binary star system with an ordinary companion star, gravity can force gas from the outer layers of the companion to spiral toward the surface of the neutron star, creating a cosmic whirlpool that astronomers call an *accretion disk*: a fast-rotating disk of hot hydrogen and helium gas (see Color Plate 14). As this gas accretes onto the surface of the neutron star, it occasionally ignites in a sudden burst of fusion. This burst is essentially a cosmic nuclear bomb, and it produces a corresponding flash of X rays. These are the X rays that we see as X-ray bursts. Because gas is continually spiraling down to the neutron star, the surface explosions tend to recur frequently—typically every few hours to every few days—which is why astronomers call such systems X-ray bursters.

Astronomers did not know precisely how a neutron star in a binary system might produce gamma rays, but it seemed fairly "obvious" that they did. Nevertheless, science requires continual testing, and astronomers therefore designed an observatory to test this assumption. Launched in 1991, the Compton Gamma Ray Observatory far exceeded the capabilities of previous gamma-ray detectors in both its sensitivity and its ability to get directional information about gamma-ray sources. In particular, one set of instruments on Compton could locate the sources of gamma-ray bursts in the sky with an angular resolution of about 1 degree—still poor compared to the resolution of less than an arcsecond routinely achieved by visible-light telescopes, but a huge improvement over previous gamma-ray detectors. Compton began detecting and locating gamma-ray bursts at a rate of about one per day. Within just a few months, Compton recorded more gamma-ray bursts than all previous gamma-ray instruments combined. It also learned enough about the locations of the burst sources to rule out the "obvious" explanation. Here's why.

If the gamma-ray bursts came from neutron stars in binary star systems, then they would have to share two crucial features with the X-ray bursts. First, because most X-ray burst sources are located in the galactic disk and not in the galactic halo, the sources of the gamma-ray bursts would also have to be concentrated in the disk. (See Mystery 7 for a review of the disk and halo.) Second, we would expect gamma-ray bursts, like X-ray bursts, to recur repeatedly in the same systems.

The all-sky map on the facing page shows the locations of gamma-ray bursts detected by Compton during its first few years of operations. Just as a flat map of the Earth represents a spherical globe, an all-sky map represents the entire celestial sphere—that is, everything we see in all directions from Earth. This particular map is oriented so that the disk of the galaxy runs along the equator, while the upper and lower regions represent what we see when we look away from the galactic disk in the sky. Careful analysis of this map confirms what your eyes probably tell you. The gamma-ray bursts come from completely random directions in the sky and not primarily from the galactic disk. Moreover, no two bursts seem to come from the same place, ruling out the possibility that gamma-ray bursts recur on the same time scales as X-ray bursts. The conclusion is clear. Gamma-ray bursts do not have the same sources as X-ray bursts.

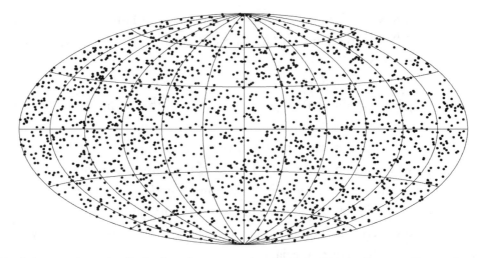

This all-sky map shows the distribution of gamma-ray bursts observed by the Compton Gamma Ray Observatory. Note that the distribution appears random, which rules out the possibility that the bursts come from sources in the galactic disk (which runs along the equator on this map).

So where do they come from? The Compton results alone could not tell us, because they gave us no information about the distances to the gamma-ray burst sources. More to the point, the Compton data allowed three general possibilities: (1) that the gamma-ray bursts came from the outskirts of our own solar system, as long as the sources were distributed spherically around the Sun (the spherical distribution would explain the random directions); (2) that they came from an extended halo of the Milky Way galaxy, as long as their typical distances were so great that our distance from the galactic center (28,000 light-years) was small by comparison (otherwise, we'd see many more bursts from one side of the galaxy than from the other); (3) that they came from objects distributed throughout the universe.

In order to select among these possibilities, astronomers needed some way to determine the distance to one or more of the gamma-ray bursts. In hopes of doing this, many teams of astronomers tried to find counterparts to a gamma-ray burst in other wavelengths. The idea was that an explosion that produced gamma rays should produce an afterglow of lower-energy light. The difficulty was that the 1-degree accuracy of the Compton observations left a fairly large piece of the sky to be searched. Moreover, the rapid nature of gamma-ray bursts required fast work. Presumably, the afterglow would fade quickly, making it difficult to see unless astronomers found it within a few hours after the gamma-ray burst occurred.

Help came with the 1996 launch of BeppoSaX, a satellite built jointly by the space agencies of Italy and the Netherlands. This satellite could locate gamma-ray bursts with an angular resolution of about 6 arcminutes and report the results to astronomers around the world within a couple of hours. Thanks to these improved data, astronomers struck the mother lode in 1997. They found the afterglow of a gamma-ray burst, clearly located in a very distant galaxy. Several more such discoveries have been made since. The case is closed—gamma-ray bursts come from distant galaxies.

However, locating the sources only deepened the mystery. The fact that gamma-ray bursts appear fairly bright to our detectors but occur very far away means that they must be enormously powerful explosions. In fact, the identified 1997 burst was described in many news reports as "the biggest bang since the Big one." This was a bit misleading both because the Big Bang was not really an explosion (see Mystery 4) and because many other gamma-ray bursts are probably comparable in power, but it made the basic point—gamma-ray bursts are the most powerful cosmic explosions ever witnessed by humans. The mystery is that we don't know what's exploding.

Before we look into some of the possible explanations for gamma-ray bursts, let's explore three key clues that tell us something about the nature of their sources. First, we know that whatever produces a gamma-ray burst must be remarkably small—far smaller in size than an ordinary star. We know this because the time scales involved in the bursts set a limit on the size of the bursting region. To understand how the time scales are related to size, imagine that you are a master of the universe and want to send a signal to a fellow master a few million light-years away. You decide to send the signal in Morse code by making some object flash on and off. But suppose you try to use an object that is one light-day across as your beacon. (One light-day is the distance light travels in one day, or about 16 billion miles.) This would make signaling a slow process, because each time you flashed the object on, the light from the front end would reach your fellow master a full day before the light from the back end. Thus, if you flashed it on and off more than once a day, your signal would be smeared out. Thinking of this idea in reverse, we see that an object that significantly brightens or dims in less than a day cannot be more than one light-day across.

This relationship between light variability and size is very important in astronomy. For example, it provides the primary evidence that the emitting regions of quasars are small—no more than a few light-hours across—which helped lead to the idea that they are powered by supermassive black holes (see Mystery 7). Flashes of light from gamma-ray bursts sometimes last only a few thousandths of a second. Because light can travel only a few hundred miles in such a short time, the source regions for these bursts can be only a few hundred miles wide. We are therefore looking for an object the size of an asteroid that, for a short time, can outshine a quasar.

A second clue about the sources comes from study of the few gamma-ray bursts for which astronomers have now detected an afterglow. Without going into details, the characteristics of the observed light allow astronomers to model how the light is probably produced. It is thought to come from shock waves generated as material moving at incredibly high speeds crashes into surrounding gas. If this conclusion is correct, the sources of gamma-ray bursts must be sending at least some matter flying outward at speeds very close to the speed of light. This idea raises the possibility that the matter may fly out only in particular directions, in which case the radiation might be "beamed" like the beam of a flashlight rather than going out in all directions like the light from a star. In that case, the gamma-ray bursts might not be quite as bright as we've thought, because calculations of their intrinsic brightnesses usually assume that they radiate in all directions. If the radiation from a gamma-ray burst is beamed along a narrow path that illuminates, say, only a hundredth of the sky, then the bursts are actually only a hundredth as bright as we've assumed—though still as bright as a quadrillion Suns. We do not yet know for sure whether the gamma-ray bursts are beamed, but we are reasonably certain that the explosions somehow propel material outward.

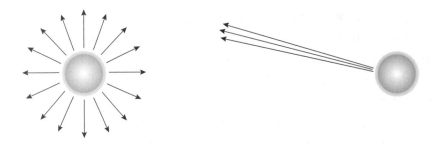

uniform radiation   beamed radiation

(Left) Objects like stars send out radiation uniformly in all directions. A small fraction of their total radiation heads in the direction of Earth. (Right) Objects that beam radiation in particular directions can be seen only if the beam points toward Earth. If gamma-ray bursts are beamed, then there may be many that we cannot see because the beam does not point to Earth.

The third clue comes from the rate at which we detect gamma-ray bursts, which is about once a day. We have good reason to believe that we are detecting nearly all the bursts that send radiation in Earth's direction, because the intensities of most recorded gamma-ray bursts lie well above current instrument detection thresholds. If many dimmer gamma-ray bursts were occurring, our instruments would be able to detect them. The fact that our detectors rarely record such dim bursts implies that there are few to be found. However, if gamma-ray bursts are beamed, there may be many that we cannot see because the beams do not point toward Earth. Thus, if the radiation is *not* beamed, we are looking for events that occur somewhere in the universe about once a day on average. If the radiation *is* beamed, we are looking for events that occur somewhat more often; depending on the width of the beam, the actual event rate might be anywhere from a few to a few thousand events per day.

These three clues significantly constrain any ideas about the causes of gamma-ray bursts. Indeed, theorists who usually can come up with dozens of viable ways to explain a new observation have been largely stymied by gamma-ray bursts. Today only two ideas are taken seriously.

The first is that gamma-ray bursts occur when two neutron stars collide with each other. This idea comes in part from the 1974 discovery of a binary star system in which both stars are neutron stars. At the time, scientists were excited primarily because this discovery suggested a way to test a central prediction of Einstein's general theory of relativity (described further in Mystery 4)—namely, that rapidly orbiting objects should emit energy in the form of gravitational waves. Ordinary stars cannot orbit each other rapidly enough for this effect to be noticeable, because their sheer size keeps their centers fairly far apart. In contrast, neutron stars are so small that they can orbit with only tens or hundreds of miles between their centers, and therefore at very high speeds. Scientists quickly calculated that the emission of gravitational waves should cause the orbits in the neutron-star binary system to decay noticeably within just a few years, and they set about observing the system to watch for this decay. The result was a stunning success for Einstein's theory: the observations precisely matched the predicted orbital decay. Thus, even though no one has yet succeeded in directly detecting gravitational waves, scientists are convinced that they exist. (New attempts at direct detection of gravitational waves will be undertaken jointly by a worldwide network of observatories in the coming decade; the American project, designed and built by scientists from Caltech and MIT, is known as LIGO—for Laser Interferometer Gravitational-wave Observatory—and has installations based in Louisiana and Washington State.)

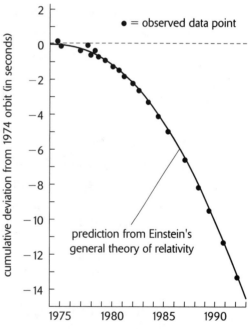

The dots on this graph show the observed orbital decay of a neutron-star binary system discovered in 1974. The solid curve shows the decay expected if the system loses energy by emitting gravitational waves, as predicted by Einstein's general theory of relativity. The match between observation and theory confirms the existence of gravitational waves. The orbital decay means that the two neutron stars will someday collide, perhaps producing a gamma-ray burst.

The relevance of the neutron-star binary system to gamma-ray bursts comes from the fact that orbital decay will inevitably cause the two neutron stars to collide someday. No one knows exactly what happens in such a collision, but it would certainly release more than enough energy to account for the power of a gamma-ray burst, and the small size of neutron stars means that the energy would be released from a very small space. Moreover, by estimating the number of neutron-star binaries in the universe, astronomers have estimated that such collisions should occur somewhere in the observable universe at least once a day, which agrees with the observed frequency of gamma-ray bursts. Of course, this agreement might just be coincidental, and the idea of neutron-star mergers as the source of gamma-ray bursts suffers from at least one major problem: despite years of effort, no one has figured out how a neutron-star collision could produce the observed characteristics of a gamma-ray burst. Nevertheless, some astronomers still favor this idea.

A more recent hypothesis involves supernovae, the cataclysmic explosions of entire stars. This idea arose when astronomers were able to study the afterglow of a 1998 gamma-ray burst in some detail. The gradual decline in the afterglow's brightness appeared to match the characteristics of an unusual type of supernova that astronomers designate as Type Ic. By mid-1999, astronomers had found similar results in the afterglows from two more gamma-ray bursts, making it highly unlikely that the correspondence between supernovae and gamma-ray bursts is coincidental.

To understand supernovae, we must recognize that stars live their lives in a constant state of battle against gravity. Because a star is essentially a giant ball of gas, its own gravity always threatens to collapse it. Fortunately for the star, during most of its life the fusion in its core generates enough internal gas pressure to counterbalance the constant gravitational squeeze. However, a star is born with only a limited supply of fuel for nuclear fusion, and when this fuel runs out gravity will cause the star to implode. If the star is relatively small, like the Sun, the implosion stops when the star becomes compressed to about a millionth of its original volume, at which point it becomes what we call a *white dwarf.* These exotic objects are essentially made of a thick plasma of bare atomic nuclei and free electrons. The pressure that halts the gravitational collapse comes from an odd effect (a quantum mechanical effect known as degeneracy pressure) that arises when huge numbers of electrons are forced to share a relatively small volume. Thus white dwarfs are the remains of relatively small stars that end their lives as gently as stars can and without any kind of supernova. (However, a white dwarf can later explode in a supernova if it gains too much mass from a companion star—see Mystery 3.)

A more massive star implodes more violently at the end of its life. As it collapses, its gravity becomes too strong to be stopped by the pressure among electrons that would make a white dwarf. The core of such a star continues to implode, forcing electrons to combine with protons to make neutrons. This process produces prodigious numbers of neutrinos (see Mystery 9)—so many that these usually ghostlike particles create a blast wave that pushes outward from the stellar core and blows the star's upper layers to smithereens. The entire process, from implosion to explosion, occurs in less than a second. The resulting light is what we see as a supernova. This process explains the origin of neutron stars: they are the balls of neutrons left behind by the core

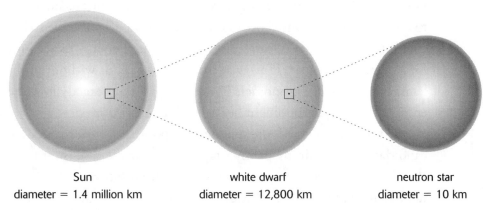

Sun	white dwarf	neutron star
diameter = 1.4 million km	diameter = 12,800 km	diameter = 10 km

The relative sizes of the Sun, a white dwarf, and a neutron star. A typical white dwarf contains the mass of the Sun in a volume the size of the Earth, so that a teaspoon of white dwarf material would weigh as much as a sport utility vehicle. A typical neutron star has up to about twice the mass of the Sun compressed into a volume only a few miles across, making a teaspoon of neutron star material as heavy as Mount Everest.

implosion. However, if the star is sufficiently massive—more than about twenty times the mass of the Sun—not even the pressure generated by neutrons pressing against one another can halt the crush of gravity. In that case, the core implosion continues until it makes a *black hole*—a place where the mass of a star is compressed to infinite density, tearing a hole in the very fabric of space. Nothing that enters a black hole can ever escape, and a black hole's boundary (called an event horizon) marks the point of no return for an object falling inward.

The supernovae associated with gamma-ray bursts appear to involve very massive stars, perhaps with as much as a hundred times the mass of our Sun. Such stars shine so brightly during their lives that they blow off their outer layers of hydrogen gas long before they die in a supernova explosion. As a result, the supernova occurs in what is essentially a bare stellar core, which allows astronomers to identify this type of supernova by its peculiar spectrum. In summary, this scenario envisions gamma-ray bursts coming from energy released during the supernova of a very massive star, in which the stellar core implodes into a black hole.

The supernova scenario is appealing not only because of the observational evidence associating gamma-ray bursts with supernovae, but also because it offers a way to explain the tremendous power of the bursts. Calculations show that the implosion of a massive stellar core into a black hole releases more than enough energy to account for a gamma-ray burst. However, this scenario is not without problems. For example, astronomers have not figured out exactly how the energy of the implosion might be turned into gamma rays.

All things considered, the supernova hypothesis is now the leading candidate for explaining gamma-ray bursts, but the case is far from closed. Even if some gamma-ray bursts come from supernovae, we cannot be sure that they all do. Clearly, we need to study many more cases in which we can find counterparts to gamma-ray bursts in other wavelengths and study the afterglows from these bursts. Astronomers are at work on several systems designed to do just that.

One such system, already operational, links space-based gamma-ray detectors with a ground-based telescope called ROTSE (for Robotic

Optical Transient Source Experiment). As soon as a space-based detector records a gamma-ray burst, a NASA network radios the approximate location of the burst to ROTSE, which then looks to the correct region of the sky. If all goes well, ROTSE can be photographing the scene in visible light within about twenty seconds after the start of the burst. If ROTSE finds a culprit in visible light, astronomers can then use more powerful telescopes to study the afterglow over the subsequent hours, days, or weeks.

Several new space missions should greatly improve our ability to detect and locate gamma-ray bursts. NASA's High Energy Transient Explorer (HETE) should already be operating by the time you read this book. HETE will locate gamma-ray bursts to within a fraction of 1 degree in the sky, making it easier for ROTSE and other ground-based telescopes to quickly look for optical counterparts. An even more ambitious NASA mission, called Swift, is being developed for launch as early as 2003. Swift will carry a gamma-ray-burst detector, an X-ray telescope, and an optical (visible-light) telescope. As soon as the gamma-ray instrument detects a burst, the spacecraft will "swiftly" rotate to point both the X-ray and optical telescopes at the source of the burst. In most cases, Swift should be obtaining X-ray and visible-light data from the burst's afterglow within a minute after the burst begins. Besides providing valuable information about the nature of the burst source and its afterglow, the Swift visible-light data should also enable astronomers to measure a redshift for the source, thereby telling us its distance.

With luck, astronomers will soon be able to study the afterglows of many gamma-ray bursts. If they all prove to be associated with the supernovae of very massive stars, then the mystery of the sources of gamma-ray bursts will be solved. Of course, just *how* the supernovae produce the gamma rays may remain mysterious for a much longer time. But it is also possible that not all the gamma-ray bursts will prove to be associated with supernovae, in which case we may have to look again at neutron-star mergers or come up with entirely new ways to explain the most powerful explosions in the universe.

*The evolution of the Universe can be
compared to a display of fireworks that
has just ended. Some few wisps, ashes and
smoke. Standing on a well-chilled cinder,
we see the slow fading of the suns and
try to recall the vanished brilliance of the
origin of the worlds.*

GEORGES LEMAÎTRE (1894–1966)

Did the Universe Have a Bout of Inflation?

The Big Bang theory is widely accepted today because it successfully predicts much about the present nature of the universe. Nevertheless, this successful theory still leaves several loose ends that so far can be explained only by invoking an esoteric idea called *inflation*—a sudden and dramatic expansion of the universe during the first tiny fraction of a second of creation. In this mystery, we will explore the underpinnings of the Big Bang theory and the loose ends tied up by inflation. In the process, we will learn how Einstein's general theory of relativity changed our view of the universe and how our view may change again as we learn whether the universe really had a bout of inflation.

Albert Einstein made important contributions to physics throughout his career, but he became a household name only after an extraordinary and successful test of his general theory of relativity. Despite its odd name, general relativity is really a theory of gravity. Prior to Einstein, the prevailing view of gravity was that of Isaac Newton, whose law of gravity says that any two masses will attract each other with a force that depends only on their masses and the distance between their centers. Newton's law of gravity works so well that it is still used for almost all situations that arise in physics, including the calculation of spacecraft trajectories and the detection of planets around other stars. On a philosophical level, however, even Newton himself objected to the law's implication of a magical "action at a distance," in which farflung objects feel each other's gravity for no apparent reason. In 1692, he wrote, "That one body may act upon another at a distance through a vacuum, . . . and force may be conveyed from one to another, is to me so great an absurdity, that I believe no man, who has . . . a competent faculty in thinking, can ever fall into it."

General relativity replaces "action at a distance" with a much more natural explanation for gravity. For the average person, the idea that general relativity is in any way "natural" might seem strange, especially since it is reputed to be a difficult and complex theory. However, this reputation really applies only to the mathematics involved in its calculations. If you focus on the underlying ideas, the theory is actually quite simple—as long as you're willing to think about more than the usual three dimensions of space (i.e., forward-back, left-right, and up-down). In particular, general relativity asks us to think of the universe as a "fabric" of intertwined space and time—or *spacetime,* as it is usually called. We cannot truly visualize this fabric, because it encompasses all three dimensions in which we live, but we can represent it by analogy with a stretched rubber sheet like a trampoline. General relativity says that a large mass, such as a planet, star, or galaxy, distorts the fabric of spacetime in a manner similar to the distortion that occurs if you place a heavy weight on a trampoline. This distortion leads to all the effects of gravity, which we can see by extending the analogy.

If you place a marble in the depression created by the heavy weight on the trampoline, the marble will roll toward the bottom of the depression. If you give the marble a push in any direction that is not aimed directly at the bottom, it will follow an "orbit" around the depression; in the absence of friction, it would stay on this orbit forever. There's nothing mysterious about the marble's paths. They are dictated by the shape of the depression, and we wouldn't think to attribute them to any kind of "action at a distance" emanating from the heavy weight. In the same way, general relativity says that a dropped rock falls to the Earth not because of any mysterious force coming from the Earth itself, but rather because the Earth creates a depression in the fabric of spacetime. Similarly, a planet orbits the Sun because the Sun distorts the fabric of spacetime rather like a heavy weight distorts the fabric of a trampoline.

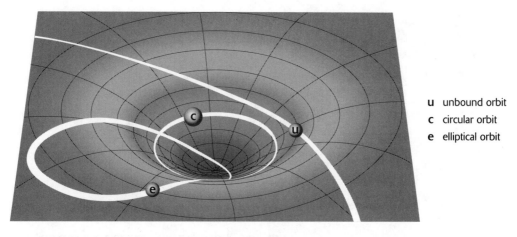

u unbound orbit
c circular orbit
e elliptical orbit

Instead of action at a distance, general relativity explains gravity as curvature of spacetime. By the trampoline analogy, masses respond to gravity much as marbles respond to a depression in the trampoline. If there were no friction between the marble and the trampoline, a marble "orbiting" about the bottom of the depression would never stop, just as planets never stop orbiting the Sun.

General relativity can be used to make specific and testable predictions because, by using its mathematical formulation, scientists can calculate precisely how much spacetime is curved by any mass. As you might expect, the results show that larger and denser masses curve spacetime to a greater extent, just as larger and denser weights create a deeper depression on a trampoline. If a mass is sufficiently great or dense, spacetime can be curved so much that the region around the mass becomes analogous to a "bottomless pit" on the trampoline. Any object that falls into a bottomless pit will keep falling forever, and there's no way out. This bottomless pit in spacetime is what we call a *black hole.* In other words, general relativity tells us that a black hole is truly a hole in the universe.

The prediction that made Einstein famous was this: stars that are seen near the Sun in the sky should appear slightly out of position. To understand this prediction, consider the path of light from a distant star. Ordinarily, we expect the starlight to come straight to Earth. This is what happens when we see the star at night, which is how we measure the star's true position in the sky. But general relativity tells us that spacetime is curved by the Sun. Thus, at times when the star appears near the Sun in the sky, the starlight passing near the Sun should

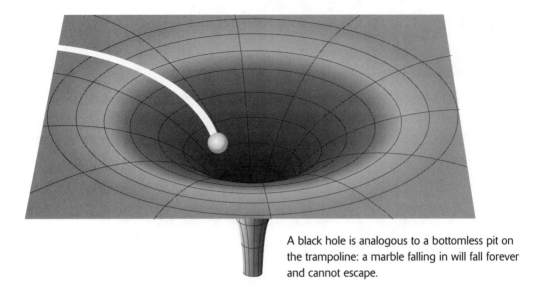

A black hole is analogous to a bottomless pit on the trampoline: a marble falling in will fall forever and cannot escape.

follow a slightly curved path, making the star appear slightly out of position. In the early 1900s, the only time that astronomers could see stars near the Sun in the sky was during a total solar eclipse. (Today it can be done almost anytime, using radio telescopes or, in some cases, space-based telescopes.) During the total eclipse of 1919, astronomers set out to test general relativity, which Einstein had published less than four years earlier. Two separate expeditions went to sites along the eclipse path—one to Principe Island off the coast of Spanish Guinea, led by British astronomer Arthur Eddington, and the other to Sobral in northern Brazil, led by Andrew Crommelin of the Greenwich Observatory. The results, announced at a joint meeting of the Royal Society and the Royal Astronomical Society on November 6, 1919, were a resounding success for Einstein, as the observed positions of stars near the occluded Sun matched the predictions of general relativity. Within weeks, Einstein's theory made headlines around much of the world.

We will have more to say about general relativity later, since its view of gravity is important to our coming discussions. For the moment, however, the theory's relevance lies in the way it changed our view of gravity from something almost mystical ("action at a distance") to

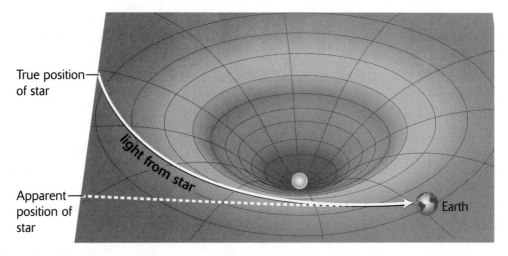

True position of star

light from star

Apparent position of star

Earth

General relativity predicts that starlight will be deflected as it passes through the curved region of space-time near the Sun, thereby making a star appear slightly out of position. This prediction was dramatically confirmed during the total solar eclipse of 1919. (The deflection in the diagram is exaggerated for clarity.)

something natural and elegant (curvature of spacetime). Einstein himself found the theory so compelling that when a student asked him how he would have felt if the eclipse observations had not agreed with his theory, he replied, "Then I would have had to pity our dear Lord. The theory is correct."

Today, there's another theory that evokes similar feelings among the scientists who have studied it in depth. Often called the *theory of inflation,* it holds that the universe underwent a sudden and dramatic surge of expansion when it was only a tiny fraction of a second old. This theory ties together a number of loose ends that would otherwise make the Big Bang more difficult to accept, and at the same time it helps unify many seemingly disparate ideas of fundamental physics. From a physicist's viewpoint, the theory is elegant and beautiful. Indeed, the theory suffers from only one significant problem—so far, there's no hard evidence to support it. Thus we have our Mystery 4. Did the universe really have a bout of inflation during the first instant of the Big Bang?

Before we get into inflation, we need to investigate why scientists find the idea of the Big Bang itself so compelling. The basic idea comes from applying simple logic to the fact that the universe is expanding, by which we mean that the average distance between galaxies (and groups of galaxies) is increasing with time. If things are moving farther apart with time, then they must have been closer together in the past. Working back, we logically conclude that the universe must have begun with everything tightly packed together, and we christen this beginning the Big Bang. Note that this does *not* make the Big Bang any kind of explosion. It just makes it a beginning. In fact, it does not even imply that the universe began from a tiny size, only that it began from incredible density. If the universe is infinite in extent (as many scientists now believe), it must have been infinite in size even when it began.

The logic leading to the Big Bang is not infallible, however, and in 1949 three respected astronomers—Hermann Bondi, Thomas Gold, and Fred Hoyle—put forth an alternative theory called the Steady State. This clever alternative managed to explain the apparent fact of universal expansion without making this fact imply a beginning. It did so by postulating that the universe is infinitely old and has always been expanding, but the ever growing gaps between old galaxies are constantly being filled in with new galaxies. Thus a Steady State universe has no beginning and no end and always looks about the same.

If you recall your high school science, you might immediately object to the Steady State theory on the grounds that it violates the law of conservation of matter and energy—and you'd be right. This law holds that the sum total of matter and energy in the universe never changes, and it is one of the most important laws in science. The Steady State theory violates this law, because it says that new matter for new galaxies is constantly being produced from "nowhere." However, calculations showed that the rate at which the Steady State theory required the production of new matter was so small that it would be virtually unnoticeable, which allowed proponents to argue that our "law" of conservation of matter and energy is really only an approximation. More important, Steady State proponents pointed out that the Big Bang, too, appears to violate the law of conservation of matter and energy. It's just that the Big Bang violates the law in one fell swoop

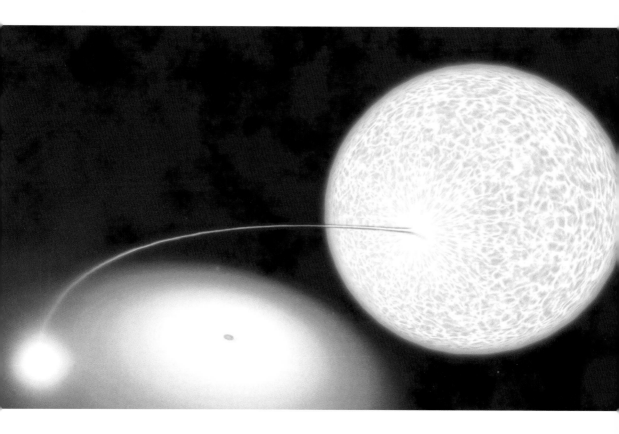

14

Artist's conception of a binary star system that produces X-ray bursts. Matter from an ordinary star spirals toward the neutron star's surface, creating an accretion disk; the neutron star resides in the center of the disk. X-ray bursts occur when some of the gas that has landed on the neutron star ignites like a cosmic nuclear bomb (not shown). Gamma-ray bursts were once thought to come from similar systems—but they don't. (Painting by Joe Bergeron.)

15

This photograph shows the Crab Nebula—the remains of a massive star whose supernova explosion was witnessed on Earth in AD 1054. The explosion left a neutron star behind in the center of the nebula.

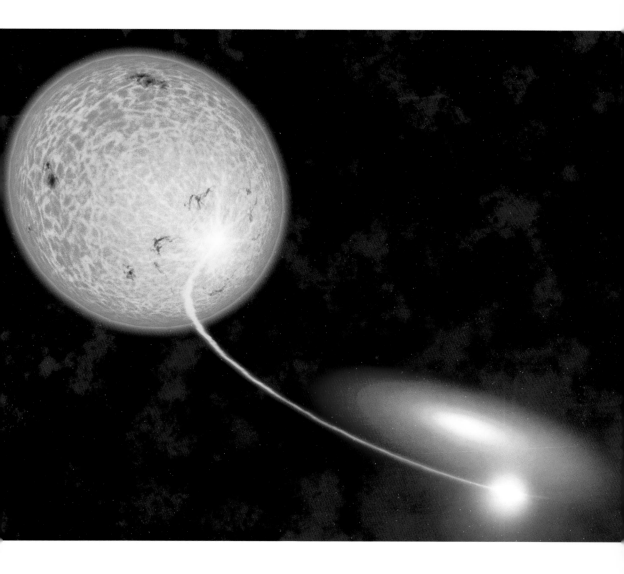

16

This artist's conception shows the type of system that may lead to a white-dwarf supernova. Gas from the still-living star spills toward the white dwarf, spiraling downward in an accretion disk. This process slowly adds mass to the white dwarf. If the white dwarf reaches a mass of 1.4 times the mass of our Sun, the laws of physics dictate that it will catastrophically explode in a supernova. Because all such white-dwarf (or Type Ia) supernovae come from white dwarfs of the same mass, they all are similar in intrinsic brightness and therefore make good standard candles. (Painting by Joe Bergeron.)

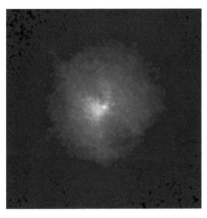

17

Visible-light (left) and X-ray (right) images of the Hydra cluster of galaxies. The X-ray image, taken by NASA's Chandra observatory, is brighter in places where more X rays are emitted. (Because X rays are invisible to our eyes, the colors in the image are false colors included to show details.) Note that X rays come from throughout the cluster, indicating that hot, X-ray-emitting gas fills the spaces between galaxies.

18

Hubble Space Telescope photograph of the galaxy cluster Abell 2218. The thin, elongated structures are distorted images of background galaxies, showing the phenomenon known as gravitational lensing. The degree of the distortions allows scientists to calculate the mass of the intervening galaxy cluster, and in some cases even to map the distribution of its dark matter.

at the moment of creation, while the Steady State theory violates it continually but gradually.

Many astronomers considered the Steady State theory to be a little too clever for its own good, but for a while, at least, its proponents often seemed ahead in the propaganda battle. Indeed, Steady State proponent Fred Hoyle, in an attempt to add some color to an appearance in a 1950 radio program, decided to ridicule the Big Bang theory by giving it a ridiculous name. This is, in fact, how the Big Bang got its name.

Nevertheless, the Steady State theory met an early death, while the Big Bang theory has survived and grown stronger with each passing decade. The reason is the same reason that always applies in science—evidence. In retrospect, the nail in the coffin of the Steady State theory was the fact that it predicted that the universe should always look about the same, whereas observations of distant galaxies now show conclusively that the universe was different in the past. (Recall that we see more distant galaxies as they were in the more distant past, so the fact that they look different from nearby galaxies means that the universe is changing with time.) However, the first crucial piece of evidence that elevated the Big Bang from a nice idea to an accepted theory was the discovery of the *cosmic microwave background,* which we discussed briefly in Mystery 7. Now it is time to look deeper.

If you've ever pumped up a bicycle tire, you know that compressing air makes it warmer, which is why the tire will feel warm to the touch when you inflate it vigorously. For the same basic reason, a "compressed universe" would have to be hotter than today's universe. The Big Bang theory holds that the universe was once compressed to incredible densities, which implies that the universe must have been very hot when it began. Thus, if there *was* a Big Bang, the universe as a whole should be filled with this remnant heat from creation.

By the early 1960s, theorists had calculated the temperatures that must have prevailed in the early universe. Then, by estimating the rate at which the universe has since cooled with time, they predicted that the universe as a whole should now have a temperature of a few degrees above absolute zero. (Absolute zero is the coldest possible temperature, which scientists call 0 Kelvin; on the Celsius scale it is –273.15°C, and on the Fahrenheit scale it is –459.67°F.) Heat is always associated with light of some kind, and at these temperatures the light should take the form of the short-wavelength radio waves called microwaves. In other words, the theorists used the Big Bang theory to predict that the universe should be everywhere filled with a background consisting of radio waves characteristic of a temperature of a few degrees above absolute zero.

The cosmic microwave background was found in 1965. As expected, it comes from everywhere in space, since it is a characteristic of the universe itself and not of any particular object within the universe. The dramatic story of how the background was discovered has been retold in many books (a personal favorite is the version in *The Red Limit,* by Timothy Ferris), so here we will focus only on the key point with regard to this discovery. The Big Bang theory predicted it, and the Steady State didn't. Score one for the Big Bang.

In fact, the Big Bang theory does much more than simply predict the existence of the cosmic microwave background. It also predicts its

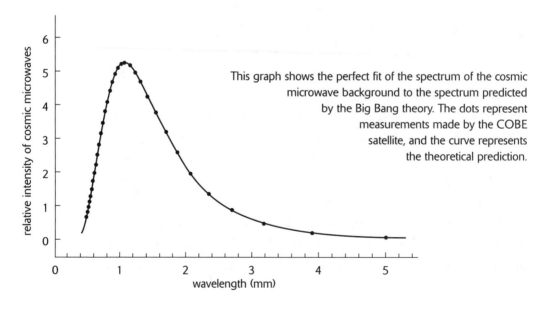

This graph shows the perfect fit of the spectrum of the cosmic microwave background to the spectrum predicted by the Big Bang theory. The dots represent measurements made by the COBE satellite, and the curve represents the theoretical prediction.

precise spectrum. NASA's COBE (COsmic Background Explorer) satellite measured the spectrum shortly after its launch in 1989, and the result was a perfect match between prediction and theory—a result that provoked a burst of applause from the audience when it was announced at a major scientific conference in January 1990. For Big Bang theorists, this was like winning the Super Bowl.

The cosmic microwave background provides very strong evidence in favor of the Big Bang, but it is not the only major evidence. A second line of evidence comes from the chemical composition of the universe. Calculations made with the Big Bang theory show that during the first second of creation the temperature was so hot that only the most fundamental particles could exist. That is, there were no molecules, no atoms, not even any ordinary atomic nuclei. Starting from these initial conditions of extremely high temperature and applying the known laws of physics, theorists can calculate how the fundamental particles should have combined with one another as the universe expanded and cooled. What they find is astonishing—and not only because they are talking about events that occurred during the first few minutes in the history of the universe!

The calculations show that the universe remained hot enough for particles to combine through nuclear fusion for only about three minutes, which means that these first three minutes determined the chemical composition of the universe for all time, except for the relatively small fraction of matter that has since undergone nuclear fusion in stars. Moreover, the calculations predict that the chemical composition of the universe should be about 75 percent hydrogen and 25 percent helium (by mass), with a trace amount of lithium and essentially nothing else. To the accuracy of current observations, once we account for the heavy elements that have been produced in stars (less than 2 percent of the total mass), this prediction of the Big Bang theory is in perfect agreement with the actual chemical composition of the universe.

The two lines of striking evidence—one from the cosmic microwave background and the other from the chemical composition of the universe—strongly suggest not only that there *was* a Big Bang but that we actually understand how it unfolded from the time the universe was a mere one second old. Thanks to this very strong evidence, the vast majority of practicing astronomers now take the Big Bang as a fact and quibble only over the details of what went on during the first second. In other words, the Big Bang itself seems to be a "fact," but the *theory* of how the Big Bang worked is still uncertain for times before the end of the first second. (The situation is analogous to that in biology, where there is broad scientific consensus that evolution is a fact in the sense that it really happened, but the theory of evolution—how it unfolds—is still a rich subject for scientific inquiry and debate.)

Despite its extraordinary successes, the "standard" Big Bang theory—that is, the theory as it stood before the idea of inflation was introduced—contains several somewhat troubling loose ends. Like Newton's idea of "action at a distance," none of these loose ends represents an actual failure of the theory. They are troubling only on a philosophical level. Let's first look at the three most important loose ends, and then come back to see how inflation ties them up.

The first loose end has to do with the presence of galaxies and large-scale structure. As we discussed in Mysteries 8 and 7, these huge structures could not exist today unless the seeds for their formation had been present in the early universe. However, the standard Big Bang theory says nothing at all about how such seeds might have come to exist. The only explanation the theory allows is that they were "already there" at the instant of creation and were not destroyed by the subsequent heat. While this may simply be the way it was, it's philosophically unsatisfying since we'd rather know how the seeds came to exist.

The second loose end is a bit more subtle but even more troubling. It has to do with the fact that the cosmic microwave background is virtually identical in all directions, which means that all parts of the universe have the same temperature. When we look at the cosmic microwave background, we are looking at light that has been traveling through space since the universe was about three hundred thousand years old. Prior to that time, the universe was so hot that any light that was emitted was quickly reabsorbed. If you could have been there, you would have been immersed in a bright fog of light, rather like what the solar interior would look like if you could be inside the Sun today. But calculations show that the universe had cooled to roughly the surface temperature of our Sun by the time it was three hundred thousand years old. Just as light escapes from the Sun's surface and flies freely through space, so was light able to fly freely through space from that time onward. Thus the light that makes up the cosmic microwave background has been traveling through space for more than 10 billion years.

This creates a loose end, because we cannot account for how the universe came to have almost precisely the same temperature everywhere. Today when we observe the cosmic microwave background in two opposite directions in the sky, we see two regions of the universe as they were when the cosmic background light was emitted, about three hundred thousand years after the Big Bang. However, these two regions are now at least 20 billion light-years apart—10 billion or more light-years in each direction, since the light has been traveling for at least 10 billion years. We can see both regions because we are right in the middle of them, but they cannot see each other because the universe is not yet old enough for light to have crossed the entire distance between them. Extrapolating the expansion backward, we similarly find that they could not have been in contact with each other at the time they emitted their microwaves. So how did the two regions end up having precisely the same temperature? We might say it's just a coincidence, but it's a coincidence as extraordinary as receiving the same letter, word for word, from two strangers in two different countries who have never met each other and who grew up in completely different, isolated cultures.

The third loose end has to do with the average density of the universe. As far as we know, gravity is the only thing that can slow the expansion of the universe, and the total strength of gravity depends on the average density of matter in the universe. If the average density is low so that the total gravity of the universe is fairly weak, then gravity will never slow the expansion by much. In that case, the universe will keep expanding forever, making what astronomers call an *open universe.* On the other hand, if the average density of matter is high, gravity should hold the universe together fairly tightly and we live in what astronomers call a *closed universe.* In that case, unless some mysterious force can counteract gravity (a possibility we will discuss in Mystery 3), the rate of expansion will gradually slow until the universe stops expanding altogether and begins to contract. Presumably, such a universe would end in a "Big Crunch." The borderline case, in which the universe teeters on the edge between open and closed, is known as a *flat universe,* for reasons we will discuss shortly. (Note: some readers may recognize a subtle distinction between "density" in general and "density of matter." You may ignore this distinction for now, but in Mystery 3 we'll explore the possibility that matter is not the only significant component of density.)

The average density in the borderline case is called the *critical density,* because it is the minimum (or critical) density that the universe must exceed if it is ever going to stop expanding. The loose end is that, according to the standard Big Bang theory, there's no reason why any particular average density should have been any more likely than any other density. Therefore, it would seem highly improbable that the universe would have begun with exactly the critical density. But current observations show that the average density of the universe today is within about a factor of 10 of the critical density (the latter can be calculated using the equations of general relativity). Moreover, calculations show that if the density today is within a factor of 10 of the critical density, it must have been within about 0.0000000000001 percent of the critical density when the universe was born. If the critical density was no more likely than any other density, why was the actual density so close to it? Again, it's possible that it "just was," but this idea is not philosophically satisfying.

Just as Einstein's general theory of relativity solved the "action at a distance" problem for gravity, the theory of inflation ties up all three of these loose ends. According to this theory, the universe did not

density of universe if there was no inflation   density of universe if there was inflation

This figure shows how the density of the universe would have changed with time depending on whether or not inflation occurred. If we simply extrapolate back in time from the current density and expansion rate of the universe, we get the smooth progression shown on the left. If inflation occurred, the very early universe would have been much denser, as shown on the right. The width of the diagram represents the volume of space into which matter is compressed. Thus, narrower means higher density on these diagrams.

expand smoothly during the earliest times. Instead, when the universe was something like a billion trillion trillionth (10^{-33}) of a second old, it underwent an incredible surge of expansion that caused the average distances between particles to grow by an enormous factor. Two pieces of the universe that were separated by less than the distance across an atom before inflation began might have been separated by a light-year when inflation ended just an instant later. So the theory of inflation predicts that during the first tiny fraction of a second, the universe was much denser than what we arrive at by simply extrapolating the current expansion backward. Let's see how this idea ties up our three loose ends.

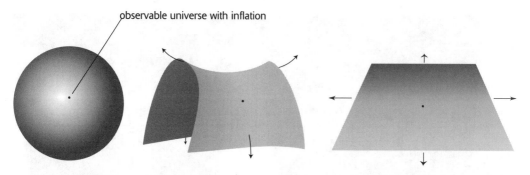

observable universe with inflation

Giant-trampoline analogies to the three possible overall shapes of the universe. The trampoline shaped into the surface of a sphere represents a closed universe, the saddle shape extending to infinity represents an open universe, and the plane represents a flat universe. The small, dark circles represent our observable universe if inflation occurred. In that case, our observable universe is so tiny compared to the scales on which any curvature is noticeable that the universe will inevitably look flat, no matter what its actual shape.

To tie up the third loose end, we must look at what general relativity tells us about the overall shape of spacetime. Earlier, when we saw that gravity acts like a depression on a trampoline, the trampoline represented only a fairly small piece of the universe, such as the region around our Sun. But what happens when we look at the *entire* universe?

According to general relativity, if the total strength of gravity is strong enough to make a closed universe, then spacetime must curve back on itself to the point at which its overall shape is analogous to that of the surface of a sphere. To extend our trampoline analogy, it's as though the trampoline becomes a huge rubber balloon. The universe expands rather like a balloon being pumped up, and the contraction that will someday follow will be like the balloon when its air is let out. The difference between the analogy and reality is that we must think of the two-dimensional balloon surface as representing all three dimensions of space. Just as a balloon has a finite surface area, the total volume of a closed universe is finite. If you are interpreting the analogy correctly, it should be clear that the universe has no center, in the same way that the surface of the Earth has no center—New York City, for example, is no more central than Paris or Beijing. The center of the sphere is not part of the universe, just as the center of the Earth is not part of a geography lesson.

An open universe, by contrast, is one in which spacetime spreads out without limit—rather like a giant trampoline in the shape of a saddle that is imagined to extend to infinity. Again, the saddle's two-dimensional surface represents all three dimensions of space. Thus an open universe has an infinite volume at all times, even as it continues to expand forever. In this case, the universe has no center because there is no center to infinity.

A flat universe, one with the critical density, is one in which spacetime has no overall curvature, analogous to an infinitely large trampoline that is stretched flat, aside from the localized depressions created by heavy masses. A flat universe also has an infinite volume—which, again, means that it has no center.

Now suppose that the universe is closed and hence began with a spherical shape in our analogy. Inflation would have made the sphere increase in size enormously. Today, the sphere would be so large that our *observable universe*—everything that we can see even in principle—is only a very tiny part of the *entire universe* represented by the sphere. Most points on the sphere's surface are far more than 10 to 16 billion light-years away from us, which means that their light has not had time to reach us in a universe that is only 10 to 16 billion years old. Just as the Earth looks flat if you look only around your own neighborhood, the universe would appear flat as far as we can see. The same would be true if inflation had occurred in an open universe. Again, our universe would seem flat because it is so small compared to the scale on which the saddle's curvature is noticeable.

If we now recall that a flat universe has the critical density, we reach the key point. Without inflation, the fact that the actual density is so close to the critical density is an amazing coincidence. With inflation, we *expect* the universe to be flat, and hence to have an average density equal to the critical density.

Next, we turn to the second loose end, or how the universe came to have the same temperature everywhere. This one is easy, given inflation. While it would still be true that opposite regions of the universe are *now* too far apart to have any contact or mixing, prior to inflation these regions would have been virtually on top of one another. In that very early time, there would have been plenty of interaction between these regions, so that they would naturally have acquired the same temperature—in the same way that all the water in a bathtub comes to the same temperature if you mix it thoroughly. We can illustrate the idea with another analogy, in which a line of chairs represents the universe and two people sitting on two particular chairs represent two regions of the universe. Before inflation, the people are sitting near each other and communicate easily. Inflation essentially dumps a few quadrillion extra chairs into the line. The two people never move, in the sense that they never leave their own chairs. But the fact that the line of chairs expands so greatly means that they are now much too far away to communicate. In other words, inflation pushes different regions of the universe far apart by making *space itself* grow, not by moving anything *through* the universe.

Finally, our first loose end, concerning the origin of the seeds around which galaxies and larger structures grew, is tied up equally easily. Here, however, we use an idea from the branch of physics known as quantum mechanics. On very tiny scales (much smaller than an atomic nucleus), it is impossible to define the idea of density precisely. As a result, there are always tiny "quantum ripples" of density on very small scales. Ordinarily, these density ripples are too small to have any noticeable effect on larger scales. But at the moment of inflation, they would suddenly have grown by the same huge factor as everything else in the universe. A density ripple that was smaller than an atomic nucleus before inflation might have become larger than our solar system after inflation. Compared to its surroundings, such a solar-system–size ripple would have enough excess density to give gravity the kick-start it needed for the later formation of galaxies and, perhaps, other large structures. Thus inflation says that the seeds for galaxies arose naturally from the amplification of tiny quantum ripples.

All in all, the way inflation ties up all three loose ends makes it a very attractive theory. Indeed, many scientists have expressed feelings about inflation similar to those Einstein expressed about general relativity—that it would be a shame if it turned out not to be true. But science is not as much about truth as it is about verifying testable predictions. On this front, inflation has a long way to go.

The major problem is that inflation has little to offer in the way of testable predictions. Theoretical calculations show that the temperature of the universe at the time of inflation would have been so high that no foreseeable technology will be able to explore such conditions. To some extent, inflation can be tested indirectly by studying the cosmic microwave background in detail. As we discussed in Mystery 7, the cosmic microwave background contains "lumps" thought to be related to the seeds of large structures. Calculations show that if inflation was responsible for those seeds, the lumps ought to be arranged in some particular patterns but not in others. NASA's Microwave Anisotropy Probe (MAP) and the European Planck mission should be able to study the patterns of lumpiness in enough detail to determine whether they are consistent with inflation. However, even if the lumps turn out to be consistent with inflation (as most astronomers expect), such evidence will probably be too weak and indirect to convince the skeptics. To put inflation on a firmer foundation, we need a more direct test.

In fact, inflation makes only one known, directly testable prediction, which is the prediction that tied up our third loose end. The universe should be flat and have precisely the critical density. Unfortunately, the idea of inflation faces trouble on this front. The trouble is that there's increasingly strong evidence that the actual density of matter is not exactly the critical density, but instead it is somewhat less than the critical density. If true, this evidence suggests that we live in an open universe, and not in the flat universe predicted by inflation. There's one way out of this dilemma: perhaps the universe is flat not because of the density of matter alone but because of some strange kind of energy built into the universe. We'll discuss the evidence concerning this strange energy in Mystery 3, but for now the idea is still speculative. Alternative proposals for explaining the "insufficient" density of the universe involve adding substantial complications to the basic idea of inflation. Because the primary appeal of inflation is its simplicity and elegance, it's hard to embrace these new ideas without solid evidence that they are correct.

Adding to the confusion is the fact that as physicists have tried to explain the source of the energy that drove inflation, their theories (the so-called *grand unified theories,* or GUTs) have made a number of other surprising predictions. One prediction is that there might be an infinite number of universes, each with its own properties and physical laws. Moreover, these theories suggest that an advanced civilization might be able to create an entire universe from nothing. Some physicists hold that such predictions make our universe more understandable and the whole of physics more "elegant." Others say just the opposite.

The bottom line is that, while inflation makes a compelling story, it remains little more than speculation at this time. If it turns out not to be true, we'll be back to dealing with the loose ends to the Big Bang theory. While this would not by itself be reason to abandon the Big Bang, it would certainly be troubling on a philosophical level.

Indeed, despite the strong evidence for the Big Bang, it is conceivable that our current ideas about the origin of the universe could turn out to be rather like Newton's idea of gravity. In retrospect, we can say that Newton's idea of "action at a distance" gave the right results for gravity but for the wrong reason. Newton could not have foreseen the correct reason at the time, because ideas of additional dimensions, the intertwining of space and time, and curved space lay too far in the future. Perhaps scientists a few hundred years from now will look back at us and say that our idea of the Big Bang gave the right results—that is, successfully predicted the cosmic microwave background and the chemical composition of the universe—for the wrong reasons. Maybe a true understanding of the origin of the universe requires ideas that no one has yet thought of.

Our quest to understand whether the universe had a bout of inflation has taken us into some of the most subtle and esoteric territory in physics, but we can now see that it is really about one of the deepest mysteries imaginable: How did the universe come to be? That is why inflation will remain a hot topic of theoretical investigation until someone comes up with a way to test the idea definitively, at which time the theory will either live or die. No one knows how long this theoretical quest may take or where it might lead once answered. It is certainly possible that our Mystery 4 will be solved early in the third millennium. It is also possible that it will survive to make the list of mysteries for the fourth millennium.

It is difficult beyond description to conceive that space can have no end; but it is more difficult to conceive an end. It is difficult beyond the power of man to conceive an eternal duration of what we call time; but it is more impossible to conceive a time when there shall be no time.

THOMAS PAINE (1737–1809)

MYSTERY
3

What Is the Fate of the Universe?

It seems that our expanding universe has two possible fates. Either the universe will continue to expand forever, or the expansion will someday stop and reverse, causing the universe to collapse in a Big Crunch. In principle, it should be easy to discover which fate awaits our universe simply by determining whether or not there is enough gravity in the universe to halt the expansion someday. But in practice this question has proved challenging. First, it is not so easy to measure the precise expansion rate and the precise strength of gravity. Second, and more important, recent observations have suggested that at least one mysterious force may be counteracting the effects of gravity on large scales. In this mystery, we will discuss the factors that determine the fate of the universe and see why this supposedly simple question may rock the foundations of much of physics and astronomy.

"Some say the world will end in fire/Some say in ice." Robert Frost's poetic line is famous in astronomy because it captures what appear to be the two possible fates of our expanding universe—reversal of the expansion, ending in a hot Big Crunch; or eternal expansion, with the universe slowly cooling and each particle of matter becoming infinitely far away from every other particle.

For the past several decades, astronomers assumed that we would learn which of these two fates awaits the universe if we could measure just two numbers precisely: the current rate of expansion and the average density of matter in the universe. The expansion rate essentially tells us how much outward "kick" the universe still has from the Big Bang, while the density of matter tells us the overall strength of gravity working to pull the universe back together. Indeed, with a bit of calculation, the two ideas can be combined in a single question: How does the actual density of the universe compare to the critical density? Recall from Mystery 4 that the critical density is the density that the universe needs if gravity is to be strong enough to someday halt its expansion. If the actual density is less than the critical density, then we live in an open universe in which the expansion will continue forever. If it is more than the critical density, then we live in a closed universe in which the expansion will someday reverse.

We do not yet know the average density of the universe, largely because most of the matter in the universe gives off very little light, making it very hard to tally. (We'll discuss this so-called *dark matter* in Mystery 2.) Nevertheless, even with fairly liberal interpretations of the evidence, it seems almost certain that the average density of matter in the universe is less than the critical density. Indeed, the density evidence now seems to point so strongly to an open universe that if this were the end of the story, the fate of the universe might barely qualify as an unsolved mystery at all. But this is not the end of the story, because several observations in the 1990s threw the long-held beliefs about what determines the fate of the universe into confusion. As a result, the fate of the universe remains a topic of contentious debate, with such deep ramifications for all of astronomy that it clearly ranks as our Mystery 3.

To understand the problems, we need to revisit how the universe expands in more depth. This will allow us to see how the expansion rate might change with time. In Mystery 8, we likened the expansion of the universe to an expanding raisin-cake, in which the raisins represent groups of galaxies. The raisin-cake model does a good job of explaining Hubble's law, or why uniform expansion means that we see the more distant galaxies moving away from us faster. But the raisin-cake model perpetrates a subtle misconception about expansion because a cake expands into a surrounding empty space. This gives the mistaken impression that the universe is a collection of galaxies expanding into a great void. In fact, the universe is not expanding *into* anything. If it were, there would be edges somewhere, beyond which there are no more galaxies, just as the raisin cake has edges beyond which there are no more raisins. According to our present understanding, this is not the case. Rather, the universe looks more or less the same no matter where you are located within it, which means that it cannot have any edges or any center. In other words, the universe encompasses *all* of space, and the universe expands because new space is created between its groups of galaxies.

Thus more realistic models of the universe must have no center and no edges. As noted in Mystery 4, the surface of a balloon can fit the bill, as can an infinite surface like that of an infinite saddle or plane (see figure on page 128). Because it's hard to visualize infinity, let's use the balloon as a model for our universe. Remember that such a model requires that we interpret the balloon's surface as representing all three dimensions of space; the inside and outside of the balloon have no meaning in this model. But aside from this mental leap, the balloon model works well because its spherical surface has no center and no edges, just as no city is the center of the Earth's surface and there are no edges where you could walk or sail off the Earth. We can represent groups of galaxies with plastic polka dots attached to the balloon, and we can make our model universe expand by blowing into the balloon. As the real universe expands, the distances between groups of galaxies grow, but the groups themselves do not grow because gravity binds their galaxies together. In a similar way, as the surface of the balloon expands, the polka dots move apart, but the dots themselves don't grow.

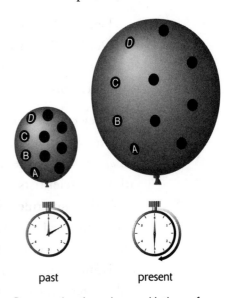

past   present

Representing the universe with the surface of a balloon. The surface has no center and no edges, making it a good model for the universe, except that we must imagine the surface to represent all three dimensions of space. When we blow into the balloon, the surface grows in size, representing expansion of the universe. The plastic polka dots represent groups of galaxies, which are held together by gravity and therefore do not grow in size as the universe expands.

Just as was the case with the raisin cake, the uniform expansion of the balloon will lead miniature scientists living on any of the polka dots to see all other dots moving away from them, with more distant dots receding faster. (If you've forgotten why, look back at the table in Mystery 8 comparing the raisin distances before and after baking.) We can make our model even more realistic by imagining that the lifetimes of the miniature scientists are so short that they cannot actually observe any visible changes in the balloon. All they can do is measure the distances and speeds of various dots. Let's arbitrarily focus on miniature scientists living on a particular dot, which we'll call dot A, and suppose that they make the following observations:

Dot B is 10 cm away and moving at 1 cm/s.
Dot C is 20 cm away and moving at 2 cm/s.
Dot D is 30 cm away and moving at 3 cm/s.
And so on . . .

They could summarize these observations as follows: *Every dot is moving away from our home dot with a speed that is 1 cm/s for each 10 cm of distance.* (Note that, because our decision to focus on dot A was arbitrary, scientists living anywhere on the balloon would find precisely the same rule.) We therefore say that the term relating the distance and speed of other dots as the balloon expands—1 cm/s of speed for each 10 cm of distance—is a constant for the entire balloon. If the miniature scientists think of their balloon as a bubble, then they might call this term the *bubble constant.*

For the miniature scientists, the bubble constant not only describes the balloon's rate of expansion at the time of their observations but also tells them something important about the balloon's history. We can see why by considering two dots that happen to be separated by 10 centimeters at present. The bubble constant's value of 1 cm/s of speed for each 10 cm of distance tells us that these two dots are currently separating at a speed of 1 cm/s. If we assume that the balloon's expansion has proceeded at a steady rate since it began, so that these two dots have always been separating at 1 cm/s, then one second ago they must have been only 9 cm apart. Similarly, two seconds ago they must have been only 8 cm apart, three seconds ago they must have been only 7 cm apart, and so on. Ten seconds ago, the two dots must have been right on top of each other. In fact, we would find the same result—dots on top of each other ten seconds ago—no matter what the present distance between any pair of dots. For example, the bubble constant tells us that dots now separated by 20 cm are moving apart at 2 cm/s, which again implies that they were on top of each other just ten seconds ago, assuming a steady rate of expansion. Thus, if the miniature scientists assume their balloon has expanded at a steady rate, they would conclude that just ten seconds ago the balloon was so small that everything was on top of everything else. They might therefore say that the expansion of their balloon began ten seconds ago with a Big Balloon Bang.

What if the balloon's rate of expansion has not been steady, but instead was faster or slower in the past? If the expansion rate was faster in the past, then pairs of dots used to be separating faster than they are now. They would thus have reached their present separations in less than ten seconds, which means that the Big Balloon Bang occurred less than ten seconds ago. On the other hand, if the expansion rate was slower in the past, then the dots would have taken more than ten seconds to reach their current separations, which means that the Big Balloon Bang occurred more than ten seconds ago.

These ideas are a bit subtle if you've never encountered them before, so let's give a brief summary of the miniature scientists' findings for the balloon universe:

- They measured the present value of the bubble constant to be 1 cm/s of speed for each 10 cm of distance. They did this by measuring both the distances and speeds of various dots moving away from their own dot.

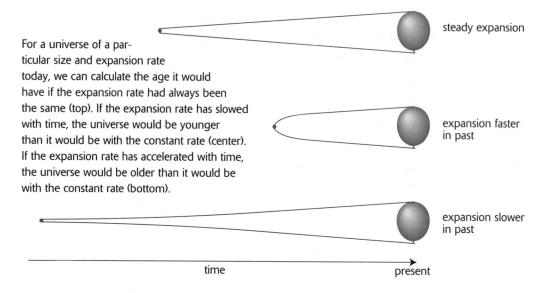

For a universe of a particular size and expansion rate today, we can calculate the age it would have if the expansion rate had always been the same (top). If the expansion rate has slowed with time, the universe would be younger than it would be with the constant rate (center). If the expansion rate has accelerated with time, the universe would be older than it would be with the constant rate (bottom).

steady expansion

expansion faster in past

expansion slower in past

time   present

- This value of the bubble constant implies that, *at present speeds,* all pairs of dots would have reached their present separations in ten seconds. Therefore:

> If the expansion rate has been steady, the balloon universe began with a Big Balloon Bang ten seconds ago.

> If the expansion rate has slowed with time, meaning that it was faster in the past, then the balloon universe began with a Big Balloon Bang *less* than ten seconds ago. This makes the balloon universe younger than it would be with the steady expansion.

> If the expansion rate has accelerated with time, meaning that it was slower in the past, then the balloon universe began with a Big Balloon Bang *more* than ten seconds ago. This makes the balloon universe older than it would be with the steady expansion.

Now, let's apply the balloon analogy to the real universe. Just as the bubble constant characterizes the balloon's present rate of expansion, some constant must characterize our universe's present rate of expansion—perhaps you've guessed that we call it the *Hubble constant.*

In the same way that the miniature scientists determined the bubble constant by measuring the distances and speeds of various dots, we determine the Hubble constant by measuring the distances and speeds of various galaxies. Because it can be difficult to measure galactic distances precisely, we do not yet know the precise value of the Hubble constant. However, it is almost certainly between 15 and 25 kilometers per second of speed for each million light-years of distance. Thus expansion causes galaxies separated by 10 million light-years to be moving apart at a speed between 150 and 250 kilometers per second, it causes galaxies separated by 100 million light-years to be moving apart at a speed between 1,500 and 2,500 kilometers per second, and so on. Because all these words are a mouthful, astronomers usually state the Hubble constant more simply as "between 15 and 25 kilometers per second per million light-years." (Many astronomers prefer to state distances with a unit called the *megaparsec,* which is equal to 3.26 million light-years; in that case, the value of the Hubble constant is between 50 and 80 kilometers per second per megaparsec.)

We can now use the Hubble constant to learn about the history of the universe in the same way that the balloon scientists used the bubble constant to learn about the history of the balloon. Let's start with the upper value in the range of the Hubble constant's possible values, which is 25 kilometers per second per million light-years. If you do the math, you'll find that objects separating at a speed of 25 kilometers per second will reach a separation of 1 million light-years in about 12 billion years. Switching to the lower value in the range, you'll find that objects separating at 15 kilometers per second will reach a separation of 1 million light-years in about 20 billion years. Thus, if the rate of expansion of our universe has always been the same, then the expansion must have begun between 12 billion and 20 billion years ago.

However, it seems almost inconceivable that the expansion rate could have been steady for all that time. The reason is gravity. Gravity acts to pull everything in the universe together and therefore tends to slow the expansion as time passes. This implies that the expansion rate should have been faster in the past, which would make the universe somewhat younger than it would be if the expansion rate had always been the same. How much younger depends on the strength of gravity, which depends on the average density of the universe. The math gets a bit complicated, but reasonable estimates of the average density—which suggest that it is less than the critical density—put the age of the universe between about 10 billion and 16 billion years. Remember that we get the younger age if the Hubble constant is near the high end of its range of 15 to 25 kilometers per second per million light-years, and the older age if it is near the low end of this range.

The logic by which we've reached these conclusions is clear, even if the details are a bit technical. But now we're at the point where we must ask whether we've told the whole story, because if we have, the story is very simple: the universe began in a Big Bang somewhere between 10 and 16 billion years ago, and has been expanding ever since. Gravity has gradually slowed the rate of expansion, but because the density of the universe is less than the critical density, the expansion will never slow enough to stop. The universe will therefore continue to expand forever.

The hints that something might be wrong with this simple story come from attempts to measure the Hubble constant more precisely. Remember that measuring the Hubble constant requires measuring both the distances of galaxies and the speeds with which they are being carried away from us by the expansion of the universe. Both measurements involve significant complications.

We can measure a galaxy's speed quite easily (from the redshift of its spectrum, as discussed in Mystery 8), but this speed is not necessarily due to expansion alone. The particular galaxy we are looking at may also be moving in response to the gravity of other large structures. Suppose, for example, that expansion causes some galaxy to move away from us at 100 kilometers per second, but that the galaxy is also being pulled by gravity toward us at 25 kilometers per second. In that case, we'd measure its speed to be 75 kilometers per second away from us—

which is significantly different from the speed due to expansion alone. Thus, until we know precisely how gravity is tugging on the galaxies we study, we cannot be certain about their speeds due to expansion.

Galactic distances are difficult to measure, because we must rely on standard-candle techniques (again, see Mystery 8) in which we *assume* that some object in a distant galaxy has the same brightness that we've measured for similar objects nearby. This assumption is not always perfectly valid, which means that distance measurements tend to be imprecise.

Fortunately, astronomers have made significant advances in tackling both problems since the early 1990s. By analyzing the speeds of thousands of galaxies, astronomers made headway in learning to separate the portion of any particular galaxy's speed due to expansion from that due to gravitational pulls toward other objects. On the distance side, more powerful telescopes—especially the Hubble Space Telescope—have allowed astronomers to use more reliable standard candles in making distance measurements.

Astronomers generally consider the most reliable standard candles to be the Cepheid variable stars, which we discussed in Mystery 7. Recall that a Cepheid periodically dims and brightens, and that the period of this variation tells us the intrinsic brightness of the Cepheid. By comparing this intrinsic brightness to the Cepheid's brightness in our sky, we can calculate its distance. Thus, by studying a Cepheid in a distant galaxy, we can measure the distance to that galaxy. Once we get reliable, Cepheid-based distances to many galaxies, we can calibrate other standard candles and thereby improve the entire cosmic distance scale. Prior to the 1990s, no telescope was powerful enough to see individual Cepheid stars in galaxies more than a few million light-years away. The Hubble Space Telescope changed that, and researchers began studying Cepheids in galaxies up to about 100 million light-years away, which they used to calibrate other types of standard candles that can be seen at even greater distances. The first announcement of results came in 1995, and generated headlines of an "age crisis" in astronomy.

The supposed crisis was this: based on their measurements of galactic distances and speeds, a research team led by Wendy Freedman of the Carnegie Observatories found that the Hubble constant must be near the high end of the possible range we cited, making the universe about 10 billion years old. (The first results suggested it might be even younger.) However, researchers who study the ages of stars claimed that the oldest stars in the Milky Way galaxy are between 12 billion and 14 billion years old. Thus the new measurement of the Hubble constant suggested that the universe was younger than its stars, which is clearly impossible. The immediate age crisis seems to have abated since 1995. Additional measurements by the same team pushed the likely age of the universe up a bit, while new research by astronomers who study stars pushed the ages of the oldest stars down a bit. Within current ranges of uncertainty, the two ages overlap—but just barely.

Meanwhile, another prominent group of astronomers—led by Allan Sandage, a protégé of Edwin Hubble and a colleague of Freedman at the Carnegie Observatories—was also reporting new measurements of the Hubble constant, but with different results. This group also used Hubble Space Telescope observations of Cepheids to calibrate their distance scale, but for distances beyond those at which Hubble can see Cepheids they relied on a relatively unusual kind of standard candle—a special kind of supernova known as Type Ia or, more colloquially, as a *white-dwarf supernova.*

Recall that a supernova involves the complete explosion of a star. According to our present understanding, these explosions can occur in two distinct ways. The first involves a massive star (a star with at least eight times the mass of our Sun) that has exhausted all its nuclear fuel. Stripped of the details, the basic idea is this: with no fuel left to generate energy and maintain internal pressure, the massive star cannot support itself against the crush of its own gravity. In an instant, the star implodes catastrophically, but the energy released by this implosion turns the tide and drives an *explosion* that blows most of the star to bits. All that remains after such a supernova is a neutron star or black hole where the center of the star used to be and a cloud of debris pushing outward into space (see Color Plate 15). Massive-star supernovae are spectacular cosmic events, but they are useless as standard candles because they do not all have the same intrinsic brightness.

The second type of supernova involves a stellar corpse known as a *white dwarf.* White dwarfs, as noted in Mystery 5, are the remains of stars that have exhausted their nuclear fuel but are not sufficiently massive to undergo a massive-star supernova. They are extremely compact objects—though not as compact as neutron stars—typically containing as much mass as the Sun in a volume no larger than that of the Earth. If you could scoop up a teaspoon of material from a white dwarf and bring it to Earth, it would weigh several tons.

If a white dwarf sits alone in space—as will our Sun when it becomes a white dwarf in about 5 billion years—then it slowly cools and fades from view over millions of years. However, some white dwarfs live in binary star systems, and in such cases gas from the companion star may spill over toward the white dwarf (see Color Plate 16). Just as a whirlpool-like accretion disk tends to form around neutron stars in binary star systems, the infalling gas from a companion can create an accretion disk around the white dwarf. The gas in the accretion disk spirals downward, eventually landing on the surface of the white dwarf. Thus a white dwarf in a binary star system can slowly gain mass. Unfortunately for the white dwarf, the laws of physics dictate that its internal structure cannot support it against gravity if it gains too much mass. Theory tells us that a white dwarf must catastrophically explode if it reaches a mass 1.4 times that of our Sun (a mass known as the *Chandrasekhar limit*), making a white-dwarf supernova.

Because this theory holds that all white-dwarf supernovae come from white dwarfs of precisely the same mass, we expect all of them to have nearly the same intrinsic brightness. Observations of white-dwarf supernovae that occur in galaxies for which we have independent ways of measuring distance confirm this idea. Sure enough, all white-dwarf supernovae are similar in intrinsic brightness, with variations of less than 25 percent from one to another. Moreover, these variations show a pattern in which the intrinsically brighter supernovae fade from view somewhat more slowly than the dimmer ones, and this allows astronomers to narrow the uncertainty in their intrinsic brightnesses to about 10 percent.

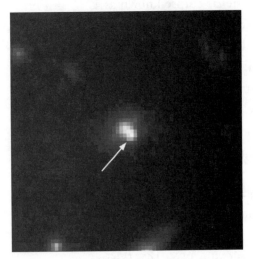

This photo shows a white-dwarf supernova in a galaxy halfway across the observable universe, recorded in March 1996 by the Hubble Space Telescope.

Both types of supernovae are rare events, typically occurring only about once a century in any particular galaxy. Indeed, humans have not witnessed a supernova explosion in our own Milky Way since 1604. Nevertheless, with billions of galaxies in the observable universe, supernovae are occurring *somewhere* at almost any instant, and by the mid-1990s several groups of astronomers had developed clever ways to find some of them. The basic technique involves comparing photographs of galaxies taken at intervals of several weeks. A supernova reveals itself as a bright spot that was not present in previous photographs. Once a supernova is discovered, the astronomers turn their telescopes to it for more detailed and more frequent study. The characteristics of its spectrum and of how its brightness changes as the explosion fades over a few weeks tell the astronomers whether it was a massive star or a white dwarf that exploded. If it is a white-dwarf supernova, astronomers can then use it as a standard candle to measure the distance to the galaxy in which it occurred.

Using white dwarf supernovae as standard candles has both an important advantage and an important disadvantage over other standard-candle techniques. The advantage is that the supernovae are extremely bright and therefore can be seen and used to measure distances to very distant galaxies. The disadvantage is that the supernovae can be used to measure distances only to those relatively few galaxies in which one happens to be caught.

Allan Sandage and other astronomers using white-dwarf supernovae to measure distances and determine the Hubble constant have thrown two more dilemmas into the mix. First, their results do not entirely agree with the results coming from Wendy Freedman's group. In particular, the results coming from the studies of the white-dwarf supernovae suggest a somewhat smaller value of the Hubble constant, implying a somewhat older universe. Indeed, when the Freedman team's studies first suggested an age crisis in the mid-1990s, the Sandage team saw no age problem at all. While the two groups no longer disagree to such an extent, the white dwarf studies still argue for somewhat lower values of the Hubble constant and hence for a somewhat older universe.

The second dilemma arising from the supernova studies is even more astonishing. Recall that the farther we look out in space, the further back we look in time (because light takes time to travel through space). Thus, when we study very distant galaxies, we are seeing them as they were long ago, when the universe was much younger. If the expansion of the universe has slowed with time—as we expect because of gravity—then these distant galaxies should appear to be moving at the higher speeds characteristic of the past, when the expansion was more rapid. But two teams of astronomers conducting supernova studies, one organized by Brian Schmidt of the Mount Stromlo and Siding Spring Observatories in Australia and the other led by Saul Perlmutter of Lawrence Berkeley National Laboratory, have found just the opposite. The more distant galaxies seem to be moving more slowly with expansion than nearby galaxies, which could be the case only if the expansion has *accelerated* with time.

The news of a possibly accelerating expansion hit astronomers like a ton of bricks, and many still are skeptical. On the positive side, an accelerating universe would alleviate any age crisis posed by the Hubble constant, since it would mean that the universe is older than a simple extrapolation of the current Hubble constant implies. On the negative side, an accelerating expansion means that some mysterious force must be counteracting gravity's tendency to slow the expansion. No known force could act this way, but there is one idea that might explain how the expansion could accelerate. This idea generally goes by the name of the *cosmological constant,* but it is also known as "Einstein's greatest blunder."

Shortly after Einstein completed his general theory of relativity in 1915, he found that it predicted that the universe could not be standing still—the mutual gravitational attraction of all the matter would make the universe collapse. However, Einstein thought at the time that the universe should be eternal and static, so he decided to alter his equations. In essence, he inserted a "fudge factor," which he called the cosmological constant, that acted as a repulsive force to counteract the attractive force of gravity. Had he not been so convinced that the universe should be standing still, he might instead have come up with the correct explanation for why the universe is not collapsing—which is because it is still expanding from the event of its birth. After Hubble discovered that the universe is expanding, Einstein called his decision to invent the cosmological constant the greatest blunder of his career.

Despite Einstein's disavowal, astronomers often still wrote the equations of general relativity with the cosmological constant included and simply assumed that its value was zero. But if the expansion is really accelerating, then the cosmological constant may not be zero after all. In that case, it represents a type of energy that fills space, gradually driving the expansion ever faster.

Here at the dawn of the millennium, the data are not yet sufficient to say definitively whether the expansion is really accelerating and there really is a cosmological constant. The case should become clear within a few years, however, as additional observations allow us to improve our measurements of the Hubble constant both for the present and for the past. If the cosmological constant proves real, at least one group of scientists will be popping champagne corks: the theorists of inflation.

As we discussed in Mystery 4, the primary difficulty with the idea of inflation is that it predicts the universe should be flat and have the critical density, while the actual density of matter in the universe appears to be less than the critical density. However, because Einstein showed that mass and energy are equivalent ($E = mc^2$), the energy embodied in the cosmological constant can also add to the total density of the universe. In other words, a universe with a cosmological constant can be flat even if the density *of matter* is less than the critical density; the density of energy makes up the difference. Perhaps you won't be surprised to learn that proponents of inflation are already arguing that the density contributed by the cosmological constant is just the amount needed to make the universe flat, thereby saving their favorite theory.

So here we stand, waiting for more observations to tell us the true value of the Hubble constant, whether the universe really is accelerating, whether there's a cosmological constant, or whether (if the different observations turn out not to be consistent with one another) we're just plain confused. The last possibility is not as flippant as it may sound, but it would mean that there must be problems in some of our most basic assumptions about the universe. For example, perhaps the Hubble constant is not the same everywhere in the universe. Or perhaps the expansion is not steadily slowing or accelerating but instead goes in fits and starts. Any of these possibilities would throw our ideas about both the birth and the fate of the universe back into confusion.

Before we leave this topic, keep in mind that the two most likely possibilities are that the measurements of an accelerating universe are in error and we live in an open universe, or that the inflation folks are right and the cosmological constant gives us a flat universe. Either way the universe will never stop expanding, and the end of history will come in what Robert Frost would have termed ice, as opposed to fire. In fact, we can describe the fate of an ever expanding universe much more precisely.

Within a few hundred billion years, all stars will burn out and die, and the now-brilliant galaxies will fade into darkness. On much longer time scales, interactions among the burned-out stars will send many of them flying into the vastness of intergalactic space, while the rest converge toward their galactic centers, merging into gigantic black holes. Here the story becomes more speculative, but if the "grand unified theories" that predict inflation are correct, the stellar corpses will eventually disintegrate into subatomic particles. Meanwhile, the giant black holes will slowly evaporate into nothingness, in a process first described by the noted British physicist Stephen Hawking. At some point in the far distant future, the universe will consist of nothing but isolated subatomic particles and isolated photons of light, each separated from every other by immense distances that will forever grow larger as the universe forever expands. Our current epoch of a universe filled with stars and galaxies will have been just a fleeting moment in an eternity of darkness.

This end in darkness may seem a bit depressing, but it is, after all, inconceivably far in the future. Nevertheless, it is fair to ask whether it is truly the end or if it could be followed by something else. Remarkably, some serious scientists already argue that there may be ways by which an intelligent civilization might survive even as the universe dies. But my personal favorite answer to this question comes from a science fiction story. In "The Last Question," Isaac Asimov begins with a couple of people asking a supercomputer whether it will ever be possible to reverse the decline of the universe, averting an end in the cold and dark. The computer responds that there are insufficient data to answer the question. Over trillions of years, computers advance and humankind survives, making the question ever more important. By the end of the story, the computer exists solely in hyperspace, outside the time and space of our universe. The universe has reached its state of ultimate darkness, with nothing left alive. Meanwhile, the computer, which Asimov calls AC, whirs on in the timelessness of hyperspace, until finally it learns how to reverse the decay of the universe:

> *For another timeless interval, AC thought how best to do this. Carefully, AC organized the program.*
>
> *The consciousness of AC encompassed all of what had once been a Universe and brooded over what was now Chaos. Step by step, it must be done.*
>
> *And AC said, "LET THERE BE LIGHT!"*
>
> *And there was light—*

Truth is indivisible, it shines with its own
transparency and does not allow itself to
be diminished by our interests or shame.

UMBERTO ECO (1932–)

What Is the Universe Made Of?

If you point a telescope almost anywhere in the sky, you'll see a field of view filled with stars and clouds of gas in our own galaxy and, behind them, distant galaxies across the universe. It would be natural to conclude that we live in a universe built primarily of stars, gas clouds, and star-filled galaxies—and totally wrong. We now have strong evidence suggesting that the vast majority of the matter in the universe—perhaps 90 percent or more—does not show up at all when we look into the sky. Moreover, we have good reason to believe that much of this so-called dark matter consists of some mysterious type of particle that we have not yet discovered. In this mystery we will discuss the evidence suggesting that the universe is made mostly of dark matter and the quest to determine the nature of dark matter.

This is how astronomers might imagine the set of instructions used by a Creator: Start with a Big Bang, run it through a fraction of a second of inflation to sprinkle it with seeds for structure, throw in a few simple laws of physics, and voilà!—10 billion years or so of cosmic evolution and out pops humanity.

The simplicity of this scenario appeals to the sensibilities of scientists, and, as we have seen, a lot of evidence backs it up, even if a few major mysteries remain outstanding. But what if something bigger is missing from this picture? What if this nice basic picture of the universe is to a large degree incomplete? It's time to turn our attention to the dirty little secret of astronomy: no one knows what the universe is made of.

Fritz Zwicky, whose work first suggested the existence of dark matter.

You may recall that several times in this book I've told you that the chemical composition of the universe is roughly 98 percent hydrogen and helium and 2 percent heavier elements. The key words here are *chemical composition,* by which astronomers mean ordinary chemical elements made from atoms. It now seems evident that such ordinary matter makes up only a small fraction—perhaps only a tenth or less—of the total mass of the universe.

The first hint that something was missing in our understanding of the universal composition came in the 1930s. The Swiss-born Caltech astronomer Fritz Zwicky, one of the more colorful astronomers of his day, was also one of the first to study clusters of galaxies. He realized that the galaxies in clusters are held together by gravity and that this gravity must cause the galaxies to swarm around the cluster's center somewhat like a swarm of bees around a beehive—except that each galaxy's orbit around the cluster center can take billions of years to complete. The stronger the overall gravity of the cluster, the faster the galaxies swarm about.

Our lifetimes are far too short for us to see the swarming, but we can still determine the speeds at which the galaxies are moving through the cluster by measuring their Doppler shifts. Measuring the average redshift of the entire cluster tells us how fast the cluster is moving away from us. Then we look at the individual redshifts of the galaxies within the cluster; those moving away from us (relative to the cluster average) have redshifts slightly larger than the cluster average, and those moving toward us have redshifts slightly smaller than the cluster average. These differences between the redshifts of individual galaxies and the redshift of the entire cluster allow astronomers to calculate the speeds of the galaxies within the cluster. From there, application of Newton's law of gravity enables us to calculate the cluster's mass.

Zwicky used this method to measure the mass of the Coma cluster of galaxies, which lies about 300 million light-years away. To his surprise, he found that the cluster was much more massive than he would have guessed just by looking at it. More specifically, by measuring the total amount of light coming from all the galaxies in the cluster, Zwicky could estimate the total number of stars that the cluster must contain in order to produce that light. He found that the actual number of stars was far short of the number needed to explain the huge mass he measured for the cluster. He therefore concluded that most of the mass in the cluster was not in the form of stars and instead must be some form that does not emit any light visible to our telescopes. Today we refer to such mass as *dark matter,* because it emits no light that we have yet been able to detect.

Zwicky's colleagues did not take his result very seriously at the time. Many assumed he must have done something wrong to reach such a strange conclusion. Others thought it might be just another of Zwicky's bizarre ideas, like his idea that supernovae are exploding stars that leave neutron stars behind—which was likewise disregarded for some three decades after Zwicky suggested it. But, starting in the 1960s, additional evidence of dark matter began to pour in.

Much of the new evidence came from the work of Carnegie Institution astronomer Vera Rubin. Women were still relatively rare in the ranks of professional astronomers, and in 1965 Rubin became the first woman permitted to observe under her own name at the Palomar Observatory, near San Diego—then the home of the world's largest telescope. (Another woman, Margaret Burbidge, had been permitted to observe at Palomar earlier but was required to apply for time under the name of her husband, also an astronomer.) During a project in which she recorded spectra of stars in the Andromeda galaxy, Rubin found results strikingly similar to those found by Zwicky some thirty years earlier.

Just as the overall strength of gravity determines how fast galaxies move in a cluster, gravity also determines how fast stars orbit within individual galaxies. Again, the basic idea is easy to understand: more gravity means that the stars feel a stronger pull toward the galactic center, and a stronger pull makes the stars move faster. Thus, by measuring the speed of any particular star or gas cloud around the galactic center, we can estimate the mass of the galaxy.

However, now we must deal with an important qualification: a star's speed around the center of a galaxy is influenced only by the portion of the galaxy's mass that lies closer to the center than the star does. If we use our own Sun's speed around the center of the Milky Way to estimate the mass of the Milky Way galaxy, we are actually measuring only the mass of the galaxy *inside* the Sun's orbit. You can see why this is true by thinking about the net gravitational force acting on the Sun. Every part of the galaxy exerts gravitational forces on the Sun, but the *net* force from matter outside the Sun's orbit is relatively small, because the pulls from opposite sides of the galaxy virtually cancel one another. In contrast, the net gravitational force from mass that lies closer to the galactic center than the Sun all pulls toward the center. (The same idea explains why you would feel no net gravitational force if you could sit at the center of the Earth. Although you would be surrounded by the Earth's mass, for every gravitational pull in one direction there would be another pull in the opposite direction, thereby leaving you weightless at the Earth's center.)

By recording spectra of stars in the Andromeda galaxy, Rubin could measure the orbital speeds of many stars from their Doppler shifts. Once she knew the orbital speed of a star, she could use Newton's law of gravity to estimate the mass of the Andromeda galaxy lying inside the star's orbit. She had expected that for stars fairly far from the galactic center the mass estimates would all be about the same—after all, pictures of the galaxy show that the vast majority of stars are concentrated in its inner regions, with relatively few stars in the outskirts. But her actual mass estimates were not at all what she'd expected. The speeds of stars in the outskirts of the galaxy suggested that there was plenty of mass out there, even though there were few stars to be seen. Just as with Zwicky's unseen mass, Rubin eventually was forced to conclude that this matter must be *dark,* in the sense of being invisible to our telescopes.

Rubin and her colleagues (most notably Kent Ford) followed up with similar studies on many other spiral galaxies. In every case, the results were essentially the same. By the early 1980s, the evidence was so overwhelming that even Rubin's harshest critics admitted that something strange was going on. Unless we are wrong in understanding how gravity works (a possibility we'll discuss later), Rubin's studies show not only that dark matter exists in spiral galaxies but that spiral galaxies consist predominantly of dark matter. Moreover, detailed investigations show that the dark matter is not mixed evenly with the luminous matter of stars. Instead, the dark matter resides primarily at large distances from the galactic center and is distributed throughout the spherical halo of the galaxy. Similar studies of elliptical galaxies likewise show them to be predominantly made of dark matter that resides at large distances from their centers.

The conclusion is astonishing: While photographs of spiral galaxies suggest that nearly all their matter resides in a disk surrounded by a nearly empty halo, the actual truth is the opposite. The stars that we see are more like the tip of an iceberg; most of the mass resides unseen in the galactic halo. In our own Milky Way, it appears that the dark matter may outweigh the luminous matter by a factor of 10 or more. In fact, we still do not know the full extent of the dark matter in our own or any other galaxy, because at very large distances from galactic centers there are no stars or gas clouds for which to measure speeds.

If galaxies themselves are built primarily of dark matter, could it be that there is also dark matter in intergalactic space? Zwicky's studies of clusters already give us at least part of the answer, which is yes. Even if we account for the dark matter in the individual galaxies within a cluster, we find that the total mass of the cluster inferred from the galaxy speeds is greater, implying that at least some of the cluster's dark matter is intergalactic. Two other types of study give the same result independently.

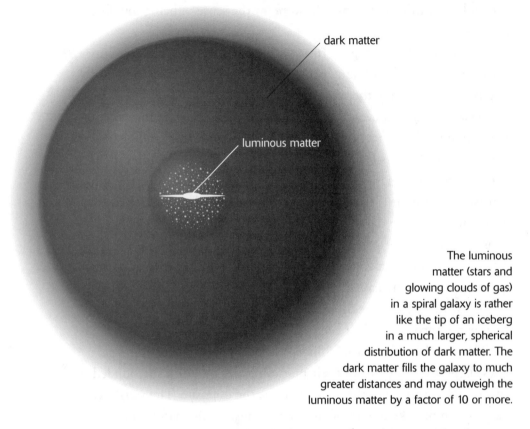

dark matter

luminous matter

The luminous matter (stars and glowing clouds of gas) in a spiral galaxy is rather like the tip of an iceberg in a much larger, spherical distribution of dark matter. The dark matter fills the galaxy to much greater distances and may outweigh the luminous matter by a factor of 10 or more.

The first of these involves looking at galaxy clusters with telescopes that observe X rays. X-ray telescopes work only in space, since Earth's atmosphere prevents cosmic X rays from penetrating to the ground. (Thankfully, or we'd all be fried!) Prior to the launch of the first X-ray telescopes in the 1960s, astronomers were unable to see gas between the galaxies in clusters. But X-ray observations showed that most clusters are filled with extremely hot but very low density gas (see Color Plate 17). The low density of this gas explains why it gives off very little visible light and hence cannot be seen with visible-light telescopes. Its high temperature—typically tens of millions of degrees—explains why it shines brightly in X rays.

Because every individual particle of gas moves in response to the gravity of the entire cluster, the speeds of the gas particles can be used to estimate the cluster mass in much the same way as can the speeds of individual galaxies. However, rather than measuring speeds of individual gas particles, astronomers measure the temperature of the gas and use the fact that temperature is a measure of the *average* speed of the gas particles. (For example, air molecules move faster on a warm day than on a cold day and therefore strike your skin harder, which is what makes you feel warm.) The masses of galaxy clusters measured by studies of X-ray–emitting gas agree with the masses found by studying speeds of individual galaxies. In addition, because the hot gas often extends into the outskirts of the clusters, these studies have shown that the dark-matter distribution continues out to great distances from the cluster center.

In case you're wondering whether this hot, X-ray–emitting gas might *be* the dark matter, the answer is no. It's true that Zwicky did not know that this gas existed when he did his studies in the 1930s, so it was "dark" to him. But even after we add all the mass of this hot gas to the masses of the individual galaxies (and the gas often outweighs the galaxies), the total still comes up well short of the total mass inferred from the measurements of galaxy speeds and gas temperature. In addition, we do not see abundant X rays coming from the halos of individual spiral galaxies, so the dark matter in these halos cannot be hot gas. Thus, when we speak of "dark matter" today, we mean that it is dark to all types of telescopes, not just to visible-light telescopes.

The second of the two other types of study involves what scientists call *gravitational lensing,* a phenomenon predicted by Einstein's general theory of relativity, and much like the bending of starlight as it passes the Sun (which we discussed in Mystery 4). Recall that starlight bends near the Sun because the Sun's gravity curves spacetime. For similar reasons, the light of a very distant galaxy that happens to lie directly behind a cluster of galaxies (as seen from Earth) will also be bent. In this case, however, the light may be bent so much that the shape of the background galaxy is highly distorted—in many cases, into a kaleidoscope of multiple images. We call this phenomenon gravitational lensing because the bending of light by gravity is somewhat analogous to the bending of light by a glass lens. Color Plate 18 shows a cluster of galaxies in which we see multiple, distorted images of more distant background galaxies. Photographs such as this one offer striking proof that spacetime really is curved in the way that Einstein suggested. They also offer astronomers a way to measure the masses of the galaxy clusters, because Einstein's equations tell us how massive a cluster must be to create a particular amount of distortion. Once again, the masses calculated in this way agree with those found by the other two methods.

Remarkably, careful study of images affected by gravitational lensing can even allow astronomers to map the distribution of dark matter in and around the cluster. Such studies are rapidly becoming more sophisticated and offer the intriguing prospect that we may someday learn how dark matter is distributed throughout the universe even before we know what it is.

In any event, given these three independent methods of weighing galaxy clusters (individual galaxy speeds, temperature of the hot intergalactic gas, and gravitational lensing), along with star speeds in individual galaxies, the evidence of dark matter is almost undeniable. The only way around this conclusion would be if we were completely wrong about how gravity works on large scales. But gravity seems to work as we expect in all other cases in which we can observe its effects on large scales and at great distances, so it is hard to imagine how we could be so wrong in our understanding of it. With the exception of a few people who continue to tinker with the theory of gravity, astronomers are near unanimous in accepting the evidence as proof that dark matter exists. The remaining questions are (1) Just how much dark matter is out there? and (2) What is dark matter made of?

Let's start with the first question. It's not an easy one to answer because the more we've looked, the more dark matter we've found. The studies of individual galaxies suggest that dark matter makes up 90 percent or more of their masses. The studies of clusters suggest that even more dark matter resides in the spaces between galaxies in clusters, and perhaps also in the outskirts of the clusters. Could there be more dark matter in superclusters or between superclusters? The answer is yes, but probably with some limits. For example, if superclusters contain huge amounts of dark matter, we ought to be able to detect the gravitational pull of this dark matter as it tugs on nearby galaxies and clusters of galaxies. Preliminary searches for such tugs have turned up some interesting hints of dark matter in superclusters, but so far the evidence does not suggest overwhelming amounts. Similarly, if there are such things as "dark clusters" containing dark matter without any luminous matter, we ought to be able to detect them by observing how they distort the images of distant galaxies by gravitational lensing. So far, there is no evidence that any dark clusters exist.

If we cut to the chase, most astronomers today will argue that dark matter outweighs the luminous matter in the universe by up to about a factor of 25 (very roughly), but not by much more. This has several important implications for our understanding of the evolution and fate of the universe. For one thing, it means that dark matter is the predominant source of gravity in the universe, which means that we must consider dark matter when we try to understand how galaxies and larger structures formed. Recall that such structures must have formed around seeds planted during the first instant of the Big Bang. As gravity began to work its magic around these seeds, dark matter must have been the main gravitational draw. Concentrations of dark matter presumably would soon have surrounded the seeds, facilitating the early growth of galaxies. Similarly, if dark matter outlines the large-scale structure of the universe—as it seems it must, since it outweighs ordinary matter—then it may still be helping the structures grow and consolidate. In that case, the strange, froth-like distributions of galaxies that we see today (see Mystery 8) may continue to change as the universe ages.

The total amount of dark matter might also have important implications for the fate of the universe. Recall that the universe will continue to expand forever unless its actual density is greater than the critical density (see Mysteries 4 and 3). The critical density is a number that astronomers calculate based on the current expansion rate, and the actual density is something that we measure. Current evidence suggests that the actual density of luminous matter in the universe is less than 1 percent of the critical density. In that case, the universe can have the critical density only if dark matter outweighs luminous matter by more than a factor of 100. If astronomers are correct in believing that dark matter outweighs luminous matter by no more than a factor of 25, then the universe does not have the critical density and will therefore continue to expand forever. (Several other lines of evidence that we have not discussed also point to the actual density being no more than about 30 percent of the critical density.)

So what is the dark matter anyway? By definition, it is any matter that we cannot detect at large distances with our current technology, and whose existence we can therefore infer only by its gravitational effects. This definition actually allows a fair range of possibilities. For example, if we placed *you* in the halo of the Milky Way, our telescopes would not be able to see you at all. Thus you are a form of dark matter, according to our definition. Of course, it's unlikely that the halos of galaxies are full of people, but dark matter could still prove to be fairly ordinary. For example, dark matter might consist of trillions of Jupiter-size "failed stars" that are too dim for us to see. Astronomers refer to objects like this as MACHOs, for MAssive Compact Halo Objects.

If the halo of our galaxy is really populated by trillions of MACHOs, they should occasionally pass directly in front of a more distant star. In that case, the gravity of the MACHO should cause a small but noticeable gravitational lensing event, in which the light of the distant star becomes distorted (magnified) when the MACHO is right in front of it. Such events have been detected, proving that at least some of the dark matter does consist of MACHOs. However, calculations show that if MACHOs accounted for all of the dark matter, we would be seeing even more gravitational lensing events than we do. Thus at least some of the dark matter must take some other form.

But what other form? It seems unlikely that it could be ordinary atoms of elements like hydrogen and helium, because such atoms should leave their marks in spectra. Moreover, an interesting result from the Big Bang theory suggests that this dark matter cannot be anything

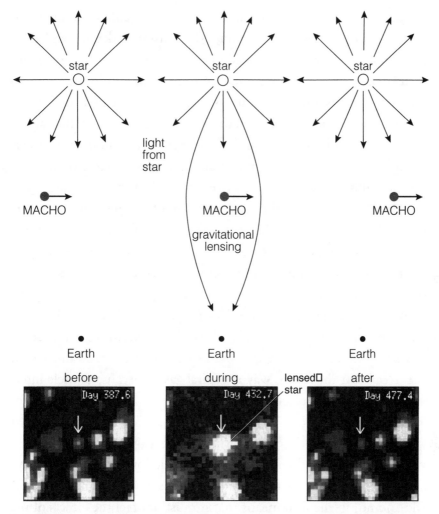

If the galactic halo is populated by trillions of MACHOs, they should occasionally cause gravitational lensing events. When a MACHO passes directly in front of a distant star along the line of sight to Earth (middle diagram), it causes bending of starlight that sends more light in our direction than usual, making the star temporarily appear brighter than normal. The inset photos show an actual gravitational lensing event; notice that the star is brighter when the MACHO is right in front of it.

made of ordinary protons, neutrons, or electrons. The details are a bit complex, but the idea is this: while fusion was occurring in the first three minutes of the Big Bang (see Mystery 4), protons and neutrons fused first into deuterium (hydrogen with a neutron in its nucleus in addition to its proton) and then into helium. Calculations show that the amount of deuterium remaining when fusion ended depended on the original density of protons and neutrons. If the density was high, all the deuterium would have fused into helium, with none left over. If the density was lower, the universe would have been left with small but noticeable amounts of deuterium. Observations show that about one out of every hundred thousand hydrogen nuclei in the universe is a deuterium nucleus (that is, it contains a neutron). Based on this observation and the calculations, theorists conclude that the density of protons and neutrons in the universe cannot be more than a few percent of the critical density. Since dark matter seems to add up to more than a few percent of the critical density, the deuterium data suggest that at least some of the dark matter is not made from ordinary protons and neutrons.

If this remaining dark matter is not "ordinary," then what is it? We have a few clues. First, it must not be made of anything that absorbs or emits much light, or else we would see it. To understand the second clue, recall that the Milky Way is a spiral galaxy because particle interactions in the protogalactic cloud caused the gas to collapse into a rotating disk (see Mystery 7). The halo of the galaxy contains stuff that got left behind when the disk formed, such as globular-cluster stars whose density made them immune to the frictional effects of the collapsing gas. Thus the fact that dark matter is located primarily in the halo and not in the disk means that it must not feel the effects of the interactions that caused the disk to form. Scientists refer to particles that do not feel such interactions as "weakly interacting," since they essentially do not interact with each other or anything else. (More technically, weakly interacting particles respond only to gravity and the weak force and not to the electromagnetic or strong forces.) We discussed one example of weakly interacting particles—neutrinos— in Mystery 9. But the third clue tells us that the dark matter in galaxies and clusters is not made of neutrinos. This clue is that the dark matter is *in* the galaxies and clusters rather than dispersed evenly throughout the universe. Because neutrinos have such tiny masses, they travel at speeds very close to the speed of light, which means they are fast enough to escape the gravitational pull of any cluster. Thus the fact that particles of dark matter are gravitationally bound in clusters means they must be slower moving than neutrinos and hence more massive.

Putting the three clues together, we conclude that the dark matter must consist of weakly interacting particles that are significantly more massive than neutrinos. Astronomers like to call them WIMPs, for Weakly Interacting Massive Particles. Aside from their being massive and weakly interacting, all other details about the nature of WIMPs remain unknown, but that doesn't stop physicists from looking for them. Like neutrinos, WIMPs should be difficult but not impossible to detect. On rare occasions they should cause noticeable interactions with ordinary matter, and if the universe is filled with them, some should be passing through the Earth at any given time. In early 2000, a group of scientists from the University of Rome created a stir by announcing that they had detected interactions that were caused by passing WIMPs. Other scientists remain skeptical of the Rome results, however, so the case for the existence of WIMPs is still far from closed. Meanwhile, several other groups are trying to detect WIMPs in the laboratory.

Before we go on, it's worth taking a moment to mention that neutrinos still have a role in dark matter—and perhaps in the fate of the universe. As we've discussed, they cannot be the WIMPs out in the halos of galaxies or in the spaces between galaxies in clusters. However, fast-moving neutrinos certainly are racing throughout the universe, and other calculations from the Big Bang theory suggest that neutrinos far outnumber ordinary particles such as protons, neutrons, and electrons. Thus, while neutrinos cannot be the dark matter in galaxies or clusters of galaxies, they still represent dark matter of another sort. The only question is how much total mass they represent, which we do not yet know because no one has successfully weighed a neutrino.

Despite the uncertainty, however, most astronomers doubt that neutrinos could represent enough total mass for the actual density of the universe to reach the critical density. The theorists of inflation who want a flat universe are probably better off placing their bets on a cosmological constant, whose energy helps the universe reach the critical density, rather than on the hope of more discoveries of massive neutrinos or other forms of dark matter.

If we step back and review where we stand, we see mind-boggling implications from the study of dark matter. The universe is made primarily of dark matter, which outweighs luminous matter by perhaps a factor of 25. Some of this dark matter may be ordinary, taking the form of MACHOs, but at least some of it is extraordinary, taking the form of WIMPs. WIMPs are particles that no one has ever identified. Thus much or most of the universe is made of something that we have not yet discovered!

Perhaps now you can see why I referred to dark matter as astronomy's dirty little secret. Like a gorilla in the basement, it's something that we try to ignore as much as possible and hope will not cause much trouble for the rest of the house. Fortunately, so far there's no reason to believe that dark matter will upset the house much. Its gravitational effects are actually helpful to the rest of astronomy because they help us explain how galaxies and large structures formed. And the fact that dark matter is made of unknown particles isn't particularly alarming, because the particles are weakly interacting and therefore shouldn't affect the rest of the universe except by their gravity.

So the general feeling among astronomers is that everything will be fine. Someday we'll identify the particles that make up dark matter, and we'll add them to our list of the constituent particles of the universe. Aside from that, we'll just continue our investigations with the foundations of physics and astronomy intact. From this viewpoint, dark matter is just a little cloud that slightly obscures our ability to learn the full story of the universe but is otherwise fairly unimportant. This is not the first time that little clouds have hung over physics and astronomy, and in most cases they have turned out to be as insignificant as they seemed. But we should close with a cautionary note.

Toward the end of the nineteenth century, some physicists thought that "the end of physics" was near, because they believed that nearly all of physics had already been discovered. Some actually encouraged their promising students to enter other fields, where there might be more opportunities for future discovery. The renowned physicist William Thomson, more commonly known as Lord Kelvin, commented that there were just a couple of nagging little problems left to solve, which he referred to as two "small clouds" hanging over physics. He enunciated both problems clearly and assumed their solutions would come quickly and easily.

In retrospect, Kelvin was right in his identification of the two major problems in physics, and they were indeed both solved within the first few decades of the twentieth century. But the solutions were not nearly as simple as he had imagined. The first problem was solved by Einstein, who showed that physicists had been mistaken in thinking that they understood space and time. The second problem was solved by Einstein, Bohr, and other pioneers of quantum mechanics, who showed that physicists had been mistaken in thinking that they understood the nature of matter and energy. The two "small clouds," in fact, turned out to be more like a London fog that had prevented scientists from appreciating some of the most basic aspects of our universe.

Could dark matter prove to be a similar fog? Could the solution to the question of what the universe is made of end up overturning large parts of our current understanding of physics and astronomy? Personally, I doubt it—but that's why it ranks as our Mystery 2.

Now when we think that each of
these stars is probably the centre of a
solar system grander than our own,
we cannot seriously take ourselves
to be the only minds in it all.

PERCIVAL LOWELL (1855–1916)

Are We Alone?

Is it possible that we are the lone intelligent beings in the universe? Or does the universe teem with civilizations, with whom we might someday join? This question probably has deeper philosophical implications than any other question in astronomy and hence is our number one mystery of the cosmos. In this final mystery, we will discuss this question and see why it has such profound implications for the future of human civilization.

Picture yourself in paradise, stretched out on a long, sandy beach under the tropical sun. As you sit back enjoying the cool breeze, the reggae music, and a frothy drink, take a moment to scoop a handful of sand. Let the sand slip *very* slowly from your palm, so that you can watch each shimmering, grain of sand as it falls. Imagine that each of those tiny grains represents an entire star system, perhaps with dozens of planets and moons, somewhere out in space. What wonders you could be holding in the palm of your hand!

Now look up and down the shoreline and imagine that every grain of sand on the long beach is a star system. That's a lot of star systems, but don't stop there. Try to imagine counting all the grains of sand on your beach . . . and then continuing on to count all the grains of sand on every beach on the entire planet Earth.

You could not possibly count so many grains of sand in a lifetime, or even in a trillion lifetimes. But if you ever somehow managed to finish, you'd discover a remarkable fact: the total number of grains of sand on all the beaches of Earth is less than the total number of stars in our observable universe.

The number of stars in the observable universe is larger than the number of grains of sand on all the beaches on Earth.

With such a myriad of stars in the universe, it's hard to imagine that ours could be the only planet where anyone is thinking about the cosmos. In this sense, it is fitting that the epigraph for this mystery comes from Percival Lowell, with whom we began in Mystery 10. Lowell was mistaken in his belief in a Martian civilization, but in his thinking he was in many ways ahead of his time. Still, *X-Files* and UFO enthusiasts notwithstanding, we have no evidence that intelligence exists anywhere else in the universe. Thus we reach our Mystery 1, which may well be the number one mystery of all human existence: Are we alone?

Before we begin to consider this question scientifically, let's take a few moments to discuss UFOs and the myths of Hollywood. Thousands of reputable citizens claim to have seen UFOs, and a substantial percentage of the public believes that we are regularly visited by aliens. If these folks are right, then our Mystery 1 is not a mystery at all.

A guiding principle of science is that the burden of proof falls on those who claim new discoveries. Even Einstein's theories were never taken at his word but took hold only after he and others supplied strong evidence. For example, as we discussed in Mystery 4, Einstein's general theory of relativity gained wide acceptance only after astronomers verified its predictions during the total eclipse of 1919. Moreover, scientific evidence must be verifiable by anyone, at least in principle; for instance, anyone can observe stars during a solar eclipse and personally verify Einstein's prediction. Generally speaking, claimed discoveries are accepted as real only if they are independently verified by others.

Given the large number of people who claim to have seen UFOs, we can be reasonably confident that most of them have seen at least *something*. But it's a huge leap from seeing "something"—which could be a weather balloon, a rocket trail, an odd cloud, or even Venus (which generates many reports of UFOs when it is particularly bright in our sky)—to genuine alien visitations. According to the principles of science, we should not accept the hypothesis of alien visits without strong evidence. So far, at least, all the publicly offered "evidence" of UFOs has been laughably weak, generally consisting of fuzzy photographs or videos, unsupportable claims that small scraps of metal could not have been produced with our technology, or first-person accounts of time spent under alien scalpels. UFO enthusiasts insist that the solid evidence is being hidden by the government, but the idea of a secret government conspiracy is hard to swallow. For one thing, it would work only if the government conspirators always got to the evidence first. Given how rapidly news cameras arrive at newsworthy events throughout the world, it's hard to believe that the government would beat the TV crews every time. And even if the government did get all the good evidence, do you really think it could be kept secret for decades? Thousands of government workers would have to have seen the evidence by now, and anyone who sneaked out a few indisputable photos would become an instant celebrity and very rich.

Perhaps because they have no strong hand to show, UFO enthusiasts often take the antiscientific approach of placing the burden of proof on those who don't believe them. You can see the problem with this approach immediately. Whereas a single piece of indisputable evidence could prove that UFOs do exist, an absence of evidence can never be used to prove that they *don't* exist. If I claim that weightless pink elephants are dancing on your head but are visible only to me, there's nothing you can do or say that will prove me wrong. In the same way, no one can prove that UFOs don't exist, but until someone provides solid proof that they do, the question "Are we alone?" remains unanswered.

With that said, we can turn back to the scientific search for other civilizations. If we want to learn whether other civilizations exist, we ought to have at least some rational, systematic way of studying the question. We therefore try to make educated guesses about what other civilizations might be like and where we might look for them. After all, with more stars than grains of sand, we could not possibly look everywhere.

A good starting point is a guess as to how many civilizations we might reasonably expect to find. To keep the numbers simpler, let's focus on our Milky Way galaxy rather than on the entire universe. This still gives us plenty of star systems to choose from. Based on star counts and the mass of the Milky Way, we know that our galaxy contains between about 100 billion and 1 trillion star systems. To get a sense of these numbers, imagine that you're having difficulty falling asleep at night, so you decide to count 100 billion star systems. How long will it take? If you want to try this, you'll need to follow a few important rules: Don't take breaks, don't fall asleep, and, most important, don't die. Because if you count at a rate of one star system per second, it will take you more than three thousand years to reach 100 billion—and more than thirty thousand years to reach 1 trillion.

In principle, we could estimate the number of civilizations in the Milky Way galaxy if we knew just a few basic facts. First, we'd need to know the number of planets in the galaxy on which life might potentially have arisen. For simplicity, let's restrict ourselves to considering Earth-like planets that have solid surfaces, oceans of liquid water, and pleasant atmospheres. Because we are going to work our way to a simple formula, let's call this number N_{EP} (for "number of Earth-like planets"). Second, we'd need to know what fraction of these Earth-like planets actually have life on them; let's use f_{life} to stand for this fraction. Multiplying $N_{EP} \times f_{life}$ would tell us the number of life-bearing planets in the Milky Way. Third, we'd need to know the fraction of the life-bearing planets on which a civilization eventually arises. Let's call this fraction f_{civ}, so that the product $N_{EP} \times f_{life} \times f_{civ}$ represents the total number of planets on which intelligent beings evolve and develop a civilization at some time in the galaxy's history. Finally, we'd need to know the fraction of these planets that have civilizations now—as opposed to millions of years in the past or future—which we'll call f_{now}. Putting all these ideas together gives us the following simple formula for the number of civilizations now inhabiting the Milky Way galaxy:

$$\text{Number of civilizations} = N_{EP} \times f_{life} \times f_{civ} \times f_{now}$$

This simple formula is a variation of an equation first expressed in 1961 by astronomer Frank Drake, one of the pioneers in efforts to search for other civilizations. It provides a useful way of organizing our thinking about the problem, because it tells us we need to know only four numbers to learn the answer. Indeed, it suffers from only one significant drawback: we don't know the value of any of its four terms!

The only term for which we can make even a reasonably educated guess is the number of Earth-like planets, N_{EP}. As we discussed in Mystery 6, our current understanding of star-system formation suggests that Earth-like planets ought to be common—at least among star systems with a single star (as opposed to binary or multiple star systems)—even though we don't yet have proof that any exist. Indeed, it's possible that many star systems have more than one Earth-like planet. Our own solar system may once have been this way. The fact that water once flowed on Mars (see Mystery 10) suggests that Mars may once have been Earth-like. All things considered, it is entirely reasonable to suppose that the average star system has one Earth-like planet, in which case there could be 100 billion or more such planets in the Milky Way galaxy.

Even assuming that our guess of 100 billion Earth-like planets is correct, the rest of the formula is more difficult. For the moment, we have no rational way to guess the fraction f_{life} of these 100 billion Earth-like planets on which life actually arose. The only evidence we have at all is the fact that life apparently arose rather quickly on Earth. The fossil record suggests that primitive life existed on Earth about 3.8 billion years ago, which is roughly the same time that planetary science suggests that the Earth would have become habitable. Prior to that time, our theory of solar system formation suggests that the Earth was subject to so many asteroid and comet impacts that any life that arose would have been quickly blotted out. Some scientists take the rapid appearance of life on Earth as evidence that life arises easily, in which case f_{life} could be 1. That is, life arises on every Earth-like planet. On the other hand, until we have solid evidence that life arose anywhere else, such as on Mars, it is also possible that Earth was somehow very lucky and f_{life} is so close to zero that life never arose on any other planet in our galaxy.

Similarly, we have little basis on which to guess the fraction f_{civ} of life-bearing planets that eventually develop a civilization. However, the fact that life flourished on Earth for almost 4 billion years before the rise of humans suggests that the development of intelligent life is much more difficult than the development of microbial life. Indeed, it was more than 3 billion years before life evolved from its simple origin into any kind of advanced plant and animal species (which came with the Cambrian explosion some 540 million years ago). The question is whether the Earth's history is typical. Roughly half the stars in the Milky Way are older than our Sun, so if 4 billion years is typical of the time required to develop a civilization, it should have happened already on about half the planets with life. On the other hand, if the Earth was "fast" and the typical time needed to develop complex species is much longer, then most life-bearing planets may be covered with nothing more advanced than bacteria.

The huge uncertainties in f_{life} and f_{civ} force us to consider only general scenarios. For example, if one in a hundred Earth-like planets ends up with a civilization, we may be newcomers among a billion civilizations that have arisen in the Milky Way. If one in a million develops a civilization, there might still have been some hundred thousand civilizations in the Milky Way galaxy alone. But if only one in a trillion Earth-like planets ends up with a civilization, then we may be the first in the Milky Way—and the only civilization within millions of light-years.

The final step in our formula is determining f_{now}, the fraction of civilizations that exist *now* as opposed to only in the past or future. Here we must decide what we mean by "civilization." Because we are dealing with the question of civilizations that we might actually contact, let's take a definition that presumes a civilization advanced enough to travel in space—and therefore advanced enough to be seeking life elsewhere through space travel in its own star system and telescopic studies of other stars. On the 4.6-billion-year-old planet Earth, the space age began only a little over forty years ago. Thus the fraction of Earth's history during which it has had a spacefaring civilization is about 40 in 4.6 billion, which is roughly 1 in 100 million.

If this fraction is typical, there would have to be some 100 million civilization-bearing planets at one time or another in the Milky Way in order to have another civilization out there now. However, we'd expect this fraction to be typical only if we are on the brink of self-destruction—after all, the fraction will grow larger for as long as our civilization survives. Thus, if civilizations are at all common, the key factor in whether any are out there now is their survivability. If most civilizations self-destruct shortly after achieving the technology for space travel, then we are almost certainly alone in the galaxy at present. But if most survive and thrive for thousands or millions of years, the Milky Way may be brimming with civilizations—and most would be far more advanced than ours.

Arguments like this have been batted around for at least a century and underlie the efforts that go under the general name of the *Search for Extraterrestrial Intelligence,* or SETI. You've probably heard about SETI, which was popularized in the wonderful movie *Contact* (based on an even better book of the same title by Carl Sagan). Most SETI efforts to date have involved using radio telescopes to search the skies for signs of artificial radio or television signals originating in other star systems, though some future efforts might involve looking for laser signals or other more advanced forms of communication.

One difficulty with present SETI efforts is that our current radio telescopes would not be sensitive enough to detect our own radio or television signals if they were coming from the distances of even the nearest stars. This suggests the rather annoying possibility that lots of civilizations out there could be listening but no one is broadcasting loud enough to be heard. Thus, if we want others to talk to us, it might be wise for us to try talking to others.

To date, we have deliberately broadcast only one loud radio transmission to the stars. The transmission consisted of a short message encoded as 0s and 1s. If members of an intelligent civilization picked up the message and laid out the 0s and 1s in a simple grid, they would see a small picture providing them with undeniable proof that they were not alone in the universe.

This message was broadcast from the Arecibo radio telescope in 1974. We have not yet heard back, and we don't expect to for a while. The message was directed toward a group of stars in our galaxy called M13, chosen because it is a globular cluster containing a few hundred thousand stars, thereby offering pretty good odds that it might harbor a civilization somewhere among them. However, M13 is more than twenty thousand light-years from Earth. Thus it will take our signal more than twenty thousand years to get there, and no answer will be forthcoming for another twenty-thousand-plus years after that. Even though we won't be holding our breath to hear back from M13, perhaps others have sent similar messages in the past, and perhaps one such message will be arriving here any day now.

The unpredictable odds of success have made SETI efforts controversial, with some arguing that we should not spend money and research time on a project for which there is such uncertainty of success and others arguing that the potential payoff—learning that we are not

This picture shows what the 1974 message broadcast from the Arecibo radio telescope, in Puerto Rico, looks like when arranged properly in a grid. Even if the recipients don't recognize the person and other information in the picture, it should still be clear that this message was deliberately made. Thus the recipients would have absolute proof that they are not alone in the universe. (The picture is turned sideways on this page.)

alone in the universe—is so great that it justifies substantial money and effort. In the 1970s, NASA chose to support a modest SETI effort to the tune of a few million dollars annually—less than a tenth of a percent of NASA's overall budget. However, a few prominent congressional opponents constantly fought these SETI allocations, and in 1993 Congress cut off SETI funding altogether. Fortunately for the scientists involved, SETI efforts resonate broadly with the public, and by 1995 radio searches were once again under way with private funding. The seed money included a million dollars each from Bill Hewlett and Dave Packard (founders of Hewlett-Packard), Intel cofounder Gordon Moore, and Microsoft cofounder Paul Allen. Today SETI efforts remain privately funded through the nonprofit SETI Institute (www.seti.org), based in Mountain View, California.

Aside from searching for signals, you might wonder about direct contact with other civilizations—the so-called "close encounters" that UFO enthusiasts believe in and that are depicted in the *Star Trek* and *Star Wars* series. Such encounters are not so easy to dismiss, even if the warp drives and hyperspace flights of the movies prove to be nothing more than science fiction. (Warp drive and hyperspace flight both postulate loopholes in Einstein's theory of relativity that violate no known laws of physics; however, there is also no evidence to suggest such technologies are possible, and some plausible arguments suggest they are not.)

Imagine that we survive and continue to explore space. With enough money and only a little advancement beyond our current technology, we could in principle build starships capable of traveling at up to a few percent of the speed of light. Although such starships would take decades to reach even the nearest stars, it is not out of the question that some people would volunteer for such trips. (If medical technology significantly lengthens human lifetimes, such trips might be much less daunting.) Within a few centuries, we could reasonably establish human colonies around dozens of nearby stars. As the colonies grew at each new location, some of the colonists might decide to set out for other star systems. In ten thousand years, our descendants would be spread among stars within a few hundred light-years of Earth. In a few million years we could have outposts scattered throughout the Milky Way, and would likely have explored nearly every star system in the galaxy. We would have become a true galactic civilization.

Now, if we take the idea that *we* could develop a galactic civilization within a few million years and combine it with the reasonable (though unproved) idea that civilizations ought to be common, we are led to an unsettling conclusion—someone else should already have created a galactic civilization. In fact, they should have done it a long time ago. For argument's sake, let's suppose that civilizations arise at the rate of about one per million stars, in which case some hundred thousand civilizations should have arisen in the Milky Way. Further, let's suppose that civilizations typically arise when their stars are 5 billion years old. Given that the galaxy is some 10 billion years old, the first of these hundred thousand civilizations would have arisen some 5 billion years ago, and others would have arisen, on average, every fifty thousand years since then. Under these assumptions, the youngest civilization besides ourselves would be some fifty thousand years ahead of us technologically, and most would be millions or billions of years ahead of us.

Thus we reach an interesting paradox: plausible arguments suggest that a galactic civilization should already exist, yet we have so far found no evidence of such a civilization. Or, as put simply by physicist Enrico Fermi in 1948, "Where are they?" (For this reason, the paradox is often called Fermi's paradox.)

There are many possible solutions to this paradox, but most fall into one of three basic categories:

1. There is no galactic civilization because civilizations are not common. Perhaps we are the first, or civilizations are so rare that many star systems, including ours, remain unexplored.
2. There is no galactic civilization because civilizations do not leave their home worlds—either because they are uninterested in interstellar travel or because they destroy themselves before achieving it.
3. There *is* a galactic civilization, but it has deliberately avoided letting us know of its existence. (*Star Trek* fans might think of this as the "Prime Directive" solution.)

If the first solution is correct, then our civilization is an astonishing achievement—something that few if any other species have ever accomplished. From this point of view, humanity becomes all the more precious, and the collapse of our civilization would be all the more tragic. Knowing this to be the case might help us learn to put petty bickering and wars behind us.

The second solution is much more terrifying. If thousands of civilizations before us have all failed to achieve interstellar travel, what hope do we have? Unless we somehow "think different" from all previous civilizations, this solution says that we will never go far in space. Given that humans have always explored when they had the opportunity, this almost inevitably leads to the conclusion that failure will come about because we destroy ourselves. Let's hope that this solution is wrong.

The third solution is perhaps the most intriguing. It says that we are newcomers on the scene of a galactic civilization that has existed for millions or billions of years before us. Perhaps this civilization is deliberately leaving us alone for the time being and will invite us to join when we prove ourselves worthy. If so, our entire species may be on the verge of beginning a journey every bit as incredible as that of a baby emerging from the womb and coming into the world.

We don't know which, if any, of these three explanations is the correct solution to the paradox. But perhaps you can now see why our Mystery 1 is no mere academic exercise and instead may have profound implications for our future. No matter what the answer turns out to be, learning it is sure to mark a turning point in the brief history of our species.

Epilogue

The known is finite, the unknown is infinite; intellectually we stand
on an islet in the midst of an illimitable ocean of inexplicability.
Our business in every generation is to reclaim a little more land.

THOMAS H. HUXLEY (1825–1895)

Picture yourself back at the mountain lake where we began this journey in the Introduction. It is now dawn, and you awaken to the sounds
of birds chirping and the fresh scent of trees in the cool mountain air.
You were up late contemplating the mysteries of the universe, but the
rising Sun helps you shake off your weariness. As you dip your hands
into the lake to wash your face, you create ripples that glitter with reflected sunlight. The sparkling points of light seem to dance in random patterns across the lake. With just a little imagination you can
see them dancing off to infinity. . . .

Your thoughts seem lost in the infinite unknown, yet as you look
at your own reflection in the water you realize how far we have come.
For what you see is in some sense not just your own face but also the
face of the universe striving to understand its own existence. It is a
face made of molecules intricately organized by biology from atoms
that came together in an interstellar cloud—atoms whose nuclei were
forged in stars from particles born in the fires of creation. It is the face
of a being that understands, at least in a rudimentary way, the 10 billion or more years of cosmic evolution that were required to make it.

You are, in essence, the universe contemplating itself. This con-
templation is no mean feat, and what you now begin to see is that we

live in a truly remarkable time. The intellectual giants of astronomy's past would be amazed by what any schoolchild can know today. Imagine being able to visit Aristarchus, telling him that he's not only been proved right about the Earth going around the Sun but that the Sun, in turn, is just one star among stars more numerous than the grains of sand on all the beaches of Earth. Imagine logging on to the Internet with Galileo, showing him the twin 10-meter Keck telescopes on Mauna Kea or the technology of the Sloan Digital Sky Survey. Picture yourself having a cup of coffee with Henrietta Leavitt, discussing how her pioneering work on Cepheid variables now lies at the foundation of the debate over whether a cosmological constant is causing the universal expansion to accelerate. What would you say to Edwin Hubble as you showed him pictures taken with the Hubble Space Telescope—and explained that it will soon be replaced by even more powerful devices?

Astronomical advances are coming at an astounding rate. It's conceivable that most or all of our ten mysteries will be solved within just a few decades. What then? What will our understanding of the universe be like in AD 2100 or AD 3000? There's no way we can answer today, but there's one thing we can say for sure: unless we do something to set back or destroy our civilization, our children and grandchildren will someday look into a mountain lake and see the reflections of faces far more enlightened than those of today's most brilliant scientists.

Which brings us to our final question: Could the study of these cosmic mysteries in any way affect the future of our civilization? On one level, the answer is undoubtedly yes. Nearly every major astronomical advance has been tied in some way to advances in technology that reverberate throughout human society. Einstein's theory of relativity, for example, was invented to help explain the universe but is now integral to the foundations of physics upon which modern technology is built. Beyond technological spin-offs, however, many people think of astronomy as an intellectual playground—a fun place to dabble but not particularly relevant to our everyday lives. I beg to differ.

Look again at the face reflected in the lake. Is it the same face you saw before you contemplated the possibilities of life in our solar system? Does it appear any different than it did before you thought about the birth of the universe or pondered the nature of dark matter? More important, when you consider the human race as a whole, do you still think of it in the same way that our ancestors did when they thought we lived at the center of the universe?

In the mid- to late 1600s, the Dutch scientist Christiaan Huygens made numerous discoveries in physics and astronomy. He made important advances in understanding pendulums, which allowed him to make much more accurate clocks than any made previously. He discovered several fundamental principles of optics, which he put to immediate use in building telescopes. Using the largest of his new telescopes—23 feet long—he discovered Saturn's moon Titan and learned that Saturn's rings nowhere touch the planet. He also made the first reasonable estimates of the distances to the stars, thereby becoming perhaps the first person to have any real understanding of the enormous scale of the heavens. In a book published not long before his death, he wrote:

> How vast those Orbs must be, and how
> inconsiderable this Earth, the Theatre upon
> which all our mighty Designs, all our
> Navigations, and all our Wars are transacted, is
> when compared to them. A very fit consideration,
> and matter of Reflection, for those Kings and
> Princes who sacrifice the Lives of so many
> People, only to flatter their Ambition in being
> Masters of some pitiful corner of this small Spot.

Clearly, Huygens thought that a little astronomical knowledge could go a long way toward changing human behavior. Whether he is proved right in this belief is a question that can be answered only by the beings that all of us see gazing back up at us from the surface of the water.

Glossary

accretion disk: A disk of material surrounding a white dwarf, neutron star, or black hole. The material in the disk orbits rapidly and eventually falls to the surface of the white dwarf or neutron star, or into the black hole.

Andromeda Galaxy (also called the **Great Galaxy in Andromeda** or **M31**): The nearest large spiral galaxy to the Milky Way.

angular resolution (of a telescope): Describes the ability of a telescope to resolve small details. More specifically, it is the smallest angular separation that two pointlike objects can have and still be seen as distinct points of light rather than as a single point of light.

apparent retrograde motion (of a planet): Refers to the apparent motion of a planet, as viewed from Earth, during the period of a few weeks or months when it moves westward relative to the stars in our sky.

arcminutes (or **minutes of arc**): One arcminute is 1/60 of 1°.

arcseconds (or **seconds of arc**): One arcsecond is 1/60 of an arcminute, or 1/3,600 of 1°.

asteroid: A relatively small and rocky object that orbits a star; asteroids are sometimes called *minor planets* because they are similar to planets but smaller.

Big Bang: The event that gave birth to the universe.

Big Crunch: If gravity ever reverses the universal expansion, the universe will someday begin to collapse and presumably end in a Big Crunch.

binary star system: A star system that contains two stars.

black hole: A bottomless pit in spacetime. Nothing can escape from within a black hole, and we can never again detect or observe an object that falls into a black hole.

blueshift: A Doppler shift in which spectral features are shifted to shorter wavelengths, caused when an object is moving toward the observer.

celestial sphere: The imaginary sphere on which objects in the sky appear to reside when observed from Earth.

Cepheid variable star: A particularly luminous type of pulsating star that follows a period–luminosity relation and hence is useful for measuring cosmic distances.

closed universe: The universe is closed if its average density is greater than the critical density, in which case spacetime must curve back on itself to the point where its overall shape is analogous to that of the surface of a sphere. In the absence of a cosmological constant, a closed universe would someday stop expanding and begin to contract.

cluster of galaxies: A collection of a few dozen or more galaxies bound together by gravity; smaller collections of galaxies are simply called *groups.*

comet: A relatively small and icy object that orbits a star.

cosmic microwave background: The remnant radiation from the Big Bang, which we detect using radio telescopes sensitive to microwaves (which are short-wavelength radio waves).

cosmic rays: Particles (such as electrons, protons, and atomic nuclei) that zip through space at close to the speed of light.

cosmological constant: The name given to a term in Einstein's equations of general relativity. If it is not zero, then it represents a type of energy that might cause the expansion of the universe to accelerate with time.

cosmological horizon: The boundary of our observable universe, beyond which we cannot see.

critical density: The precise average density for the entire universe that marks the dividing line between an open universe and a closed universe; a universe with the critical density is said to be flat.

dark matter: Matter that we infer to exist from its gravitational effects, but from which we have not detected any light; dark matter dominates the total mass of the universe.

deuterium: A form of hydrogen in which the nucleus contains a proton and a neutron, rather than only a proton (as is the case for most hydrogen nuclei).

disk (of galaxy): The portion of a spiral galaxy that looks like a disk and contains most of the stars, gas, and dust.

Doppler effect (shift): The effect that shifts the wavelengths of spectral features in objects that are moving toward (creating a blueshift) or away (creating a redshift) from the observer.

flat universe: Refers to the case in which the density of the universe is equal to the critical density; as with an open universe, expansion will continue forever in this case.

galaxy: A huge collection of anywhere from a few hundred million to more than a trillion stars, all bound together by gravity.

gamma-ray burst: A sudden burst of gamma rays coming from deep space.

general relativity: Einstein's theory that describes gravity as curvature of spacetime.

globular cluster (of stars): A spherically shaped cluster of up to a million or more stars; globular clusters are found primarily in the halos of galaxies and contain only very old stars.

Grand Unified Theories (GUTs): Theories that seek to explain many aspects of physics through a single set of ideas. (More technically, these theories seek to unify the strong, weak, and electromagnetic forces into a single force.)

gravitational lensing: The magnification or distortion (into arcs, rings, or multiple images) of an image caused by light bending as it passes near a massive object, as predicted by Einstein's general theory of relativity.

halo (of galaxy): The spherical region surrounding the disk of a spiral galaxy.

Hubble's constant: A number that expresses the current rate of expansion of the universe.

Hubble's law: Expresses the idea that the more distant galaxies are, the faster they move away from us.

inflation (of the universe): A sudden and dramatic expansion of the universe thought to have occurred when the universe was a tiny fraction of a second old.

interferometry: A telescopic technique in which two or more telescopes are used in tandem to produce much better angular resolution than the telescopes could achieve individually.

large-scale structure (of the universe): Generally refers to structure of the universe on size scales larger than that of clusters of galaxies.

light-year: The distance that light travels in one year, which is 9.46 trillion kilometers (or 5.87 trillion miles).

Local Group: The group of about thirty galaxies to which the Milky Way galaxy belongs.

Local Supercluster: The supercluster of galaxies to which the Local Group belongs.

lookback time: Refers to the amount of time since the light we see from a distant object was emitted; e.g., if an object has a lookback time of 400 million years, we are seeing it as it looked 400 million years ago.

MACHOs: Stands for *massive compact halo objects,* and represents one possible form of dark matter in which the dark objects are relatively large objects such as planet-size "failed stars."

Martian meteorites: Meteorites found on the Earth's surface that apparently were blasted off the surface of Mars.

Milky Way: Used both as the name of our galaxy and to refer to the band of light we see in the sky when we look into the plane of the Milky Way galaxy.

moon: An object that orbits a planet.

neutrino: A very lightweight particle that hardly interacts with other matter at all (i.e., a type of weakly interacting particle).

neutron star: The compact corpse of a massive star left over after a supernova; typically contains a mass comparable to the mass of the Sun in a volume just a few kilometers in radius.

nuclear fusion: The process in which two (or more) nuclei slam together to form a larger nucleus.

observable universe: The portion of the entire universe that, at least in principle, can be seen from Earth.

open universe: The universe is open if its average density is less than the critical density, in which case spacetime has an overall shape analogous to the surface of a saddle that is imagined to extend to infinity. If the universe is open, it will never stop expanding.

planet: An object that orbits a star and that, while much smaller than a star, is relatively large in size; there is no "official" minimum size for a planet, but all the nine planets in our solar system are at least 2,000 km in diameter.

protogalactic cloud: A huge, collapsing cloud of intergalactic gas from which an individual galaxy formed.

quasar: The name given to centers of galaxies that are unusually bright because of power generated as matter falls into a supermassive black hole.

redshift: A Doppler shift in which spectral features are shifted to longer wavelengths, caused when an object is moving away from the observer.

SETI (Search for Extraterrestrial Intelligence): The name given to observing projects designed to search for signs of intelligent life beyond Earth.

solar neutrino problem: Refers to the disagreement between the predicted and observed number of neutrinos coming from the Sun.

solar system (or star system): Consists of a star (sometimes more than one star) and all the objects that orbit it.

spacetime: The inseparable, four-dimensional combination of space and time.

spectral lines: Bright or dark lines that appear in an object's spectrum, which we can see when we pass the object's light through a prism-like device that spreads out the light like a rainbow.

standard candle: An object for which we have some means of knowing its intrinsic brightness, so that we can use its apparent brightness to determine its distance from Earth.

Steady State theory: A now-discredited theory that held that the universe had no beginning and looks about the same at all times.

stellar parallax: The apparent shift in the position of a nearby star (relative to more distant objects) that occurs as we view the star from different positions in the Earth's orbit of the Sun each year.

supercluster: Superclusters consist of many clusters of galaxies, groups of galaxies, and individual galaxies.

supernova: The explosion of a star.

universe: The sum total of all matter and energy.

weakly interacting particles: Particles that do not feel friction and do not emit or absorb light, and therefore rarely interact with other matter at all. (More technically, weakly interacting particles respond only to the weak force and gravity, but not to the strong or electromagnetic forces.)

white dwarf: The hot, compact corpse of a low-mass star; typically with a mass similar to the Sun compressed to a volume the size of the Earth.

WIMPs: Stands for *weakly interacting massive particles,* and represents a possible form of dark matter consisting of subatomic particles that hardly interact with other matter at all, but are more massive and slower-moving than neutrinos.

Appendix: Watching the Mysteries Unfold

This is an exciting time in the history of astronomy, as there are more people involved in research using more instruments and more missions than ever before. Today there are more than 5,000 active astronomers working in more than 100 countries around the world. More than two dozen nations have active programs for astronomy from space, and many more have ground-based efforts or share in the efforts of the space-faring nations. As a result, you can expect frequent news reports of important developments in the unraveling of our ten mysteries.

There are many ways you can watch the mysteries unfold, such as by reading astronomy news in newspapers and magazines or by reading other books about astronomy. One of the best ways to keep abreast of developments as they occur is with the world wide web. In this appendix, I list several categories of useful web sites. All of these have links from the web site for this book, on which I'll also post news about major discoveries as they occur. Thus you can use the book web site as a starting point, and therefore may want to bookmark it now:

www.astrospot.com

Key Missions for the Mysteries

The following table lists (by launch date) some of the major current and planned space missions that should shed light on many of the mysteries described in this book. For a more complete list of space missions, go to the web site for NASA's Office of Space Sciences (http://www.hq.nasa.gov/office/oss/missions).

Key Missions for the Mysteries					
Mission Name	*Launch Year*	*Lead Space Program/ Nation*	*Most Relevant to Mystery(ies)*	*Key Capabilities*	*Web Address*
Compton Gamma Ray Observatory (CGRO)	1991	NASA	5	gamma-ray observations	http://cossc.gsfc. nasa.gov/cossc
Hubble Space Telescope	1991	NASA	8, 7, 3, 2	Optical, infrared, and ultraviolet imaging and spectroscopy	http://www. stsci.edu
Mars Exploration Program	1996 + 2- year intervals	NASA	10	Includes multiple current and future missions to explore Mars	http://mars.jpl. nasa.gov
Cassini/Huygens	1997	NASA	10	Will arrive at Saturn in 2004; carries Huygens probe which will descend to Titan	http://www.jpl. nasa.gov/cassini
Far Ultraviolet Spectroscopic Explorer (FUSE)	1999	NASA	7, 3, 2	Ultraviolet spectroscopy	http://fuse. pha.jhu.edu
Chandra	1999	NASA	7, 3, 2	X-ray imaging and spectroscopy	http://chandra. harvard.edu
X-ray Multi-Mirror Mission (XMM)	1999	ESA*	7, 3, 2	X-ray spectroscopy	http://sci.esa.int/ missions/xmm

Mission Name	Launch Year	Lead Space Program/ Nation	Most Relevant to Mystery(ies)	Key Capabilities	Web Address
High Energy Transient Explorer (HETE)	2000	NASA	5	Study gamma-ray bursts	http://space.mit. edu/HETE
Microwave Anisotropy Probe (MAP)	2001	NASA	7, 4, 3	High-resolution study of the cosmic background radiation	http://map.gsfc. nasa.gov
Space Infrared Tele-scope Facility (SIRTF)	2002	NASA	7, 3, 2	Infrared observations of the cosmos.	http://sirtf. caltech.edu
Swift	2003	NASA	5	Study gamma-ray bursts and their X-ray and optical afterglows	http://swift. sonoma.edu
Europa Orbiter	2003+	NASA	10	Orbit and study Jupiter's moon Europa	http://www. jpl.nasa.gov/ ice_fire// europao.htm
Kepler	2005+	NASA	6	Search for transits by extrasolar planets	http://www. kepler.arc. nasa.gov
Space Interferometry Mission (SIM)	2006+	NASA	7, 6, 3	High-resolution imaging through interferometry	http://ngst.gsfc. nasa.gov
Planck	2007+	ESA*	7, 4	Higher-resolution study of the cosmic back-ground radiation	http://sci.esa. int/planck
Next Generation Space Telescope (NGST)	2008+	NASA	7, 6, 3	Follow-on to Hubble Space Telescope	http://ngst.gsfc. nasa.gov
Darwin	2009+	ESA*	6, 1	Search for Earth-like planets	http://ast.star.rl. ac.uk/darwin
Terrestrial Planet Finder (TPF)	2011+	NASA	6, 1	Search for Earth-like planets	http://ngst.gsfc. nasa.gov

*European Space Agency

Other Key Projects for the Mysteries

Besides the space missions listed, the following lists (alphabetically) major ground-based projects discussed in this book.

Project	Major Goal	Most Relevant to Mystery(ies)	Web Address
Laser Interferometer Gravitational-Wave Observatory (LIGO)	Search for gravitational waves	5, 4, 3, 2	http://www.ligo.caltech.edu
Robotic Optical Transient Search Experiment (ROTSE)	Look for optical counterparts of gamma-ray bursts	5	http://www.umich.edu/~rotse
Search for Extraterrestrial Intelligence (SETI)	Search for signals from other civilizations	1	http://www.seti.org
Sloan Digital Sky Survey	Study large-scale structure of universe	8, 7, 3	http://www.sdss.org
Sudbury Neutrino Observatory (SNO)	Study the solar neutrino problem	9	http://www.sno.phy.queensu.ca
Super-Kamiokande (Super-K)	Neutrino observatory	9	http://www.phys.washington.edu/~superk

More Astronomical Web Sites

The following web sites are some of my other personal favorites for astronomy information, listed in alphabetical order.

More Astronomical Web Sites		
Site	*Description*	*Web Address*
The American Association of Variable Star Observers (AAVSO)	One of the largest organizations of amateur astronomers in the world. Check this site if you are interested in serious amateur astronomy.	http://www.aavso.org
The Astronomical Society of the Pacific	An organization for both professional astronomers and the general public, devoted largely to astronomy education.	http://www.aspsky.org
Astronomy Magazine	Loaded with useful information and current news about astronomy.	http://www.kalmbach.com/astro
Astronomy Now	Web site for a leading British astronomy magazine.	http://www.astronomynow.com
Astronomy Picture of the Day	An archive of beautiful pictures, updated daily.	http://antwrp.gsfc.nasa.gov/apod
Canadian Space Agency	Home page for Canada's space program.	http://www.space.gc.ca
David Malin's Astronomical Images (at the Anglo-Australian Observatory)	Malin is world-renowned for his spectacular, true-color photographs.	http://www.aao.gov.au/images
European Space Agency (ESA)	Home page for this international agency.	http://www.esa.int
The Extrasolar Planets Encyclopedia	Information about the search for and discoveries of extrasolar planets	http://cfa-www.harvard.edu/planets
Mercury Magazine	Magazine of the Astronomical Society of the Pacific; check out the education columns!	http://www.aspsky.org/mercury.html
NASA Home Page	Learn almost anything you want about NASA.	http://www.nasa.gov
NASA Photo Gallery	Get access to almost any NASA image ever taken.	http://www.nasa.gov/gallery/photo

More Astronomical Web Sites *(continued)*

Site	*Description*	*Web Address*
The New York Times Exploring the Solar System	Recent articles and other information about the Sun and our solar system.	http://www.nytimes.com/ library/national/science/ solar-main.html
The Nine Planets (University of Arizona)	A multimedia tour of the solar system, with the option to study almost any solar system object in detail.	http://seds.lpl.arizona.edu/ nineplanets/nineplanets/ nineplanets.html
The Planetary Society	Has more than 100,000 members who are interested in planetary exploration and the search for life in the universe.	http://planetary.org
San Francisco State University Discovery of Extrasolar Planets	More information about extrasolar planets, from the Extrasolar Planet Search Team at San Francisco State University.	http://www.physics. sfsu.edu/~gmarcy/ planetsearch
SETI Institute Online	Devoted to the search for other civilizations.	http://www.seti.org
Sky and Telescope Magazine	Loaded with useful information and current news about astronomy.	http://www.skypub.com
Space Telescope Science Institute	Pictures and other information from the Hubble Space Telescope.	http://www.stsci.edu
Yahoo! Astronomy and Space News Headlines	About the most complete listing of recent astronomy news articles that you could want.	http://headlines.yahoo. com/Full_Coverage/ Science/Astronomy_ Discoveries

Acknowledgments

I have many people to thank for helping make this book possible. First and foremost is my wife, Lisa, who's supported me both emotionally and financially as I've made the transition from researcher to writer, followed by our 2-year-old son, Grant, who makes it a joy to work at home. I also thank my parents, Tam and Bob, and Grandma Flo for their long-term support and encouragement.

For many of the ideas in this book, I thank my good friends—and coauthors of my astronomy textbooks—Megan Donahue, Nick Schneider, and Mark Voit. Mark Voit, in particular, has been a major collaborator on this book; among other things, he helped me formulate the set of mysteries, he read multiple drafts of the text and provided many editorial suggestions, and he has helped me with many technical issues in which my own expertise is lacking. Many other excellent ideas and editorial suggestions came from the outstanding editor who worked on this book, Sara Lippincott.

For bringing this book from idea to publication, I thank my many friends at Addison Wesley in San Francisco. Because they are primarily a textbook publisher, it was a gamble for them to publish this book at all—it happened only because of the great efforts made by Linda Davis, Tim McKee, and Ben Roberts. Ben and his team—including Joan Marsh, Adam Black, Catherine Flack, Nancy Gee, Margot Otway, Claire Masson, and Blake Kim—have done a tremendous job in developing this book and in providing unending support.

Special thanks to Emi Koike for her work on the art and design of this book, and for suggesting its title. And special thanks also to Mary Douglas of Rogue Valley Publications for helping this book sail smoothly through production, just as she has done with my textbooks in the past.

Finally, I'd like to thank the many other people who have had a great impact on my thinking and therefore on this book. Carl Sagan was an important and early influence; although I met him only a couple of times, it was his *Cosmos* series that convinced me to pursue graduate work in astronomy. Dick McCray's support has been critical in keeping me focused on astronomical education and writing; he also reviewed portions of this book. Tom Ayres, Jeff Goldstein, and Cheri Morrow, three of my closest collaborators over the years, have greatly influenced my thinking about both astronomy and education. And to the many others with whom I've worked closely at the University of Colorado, NASA Headquarters, and elsewhere—I hope that each of you will be able to see your influence as you read this book, and hope you are pleased with the outcome.

Credits

Index

Cepheid variables, 48, *63*, 63–64, 143
Cerenkov radiation, 30
Chandra observatory, 75
Chandrasekhar limit, 145
Chlorine neutrino detectors, 26–27
Clusters (of galaxies), 3, 43
COBE (Cosmic Background Explorer), 76, 122–123
Collisions
    galaxies, 70–71, *71*, 73
    neutron stars, 109–110
    stars, 70
Coma cluster, *43*, 155
Comets, life on, 20
Compton Gamma Ray Observatory, 104–105
Computer models, 67–68, 70
Conservation of angular momentum, 66
Conservation of matter and energy, 120
Constellations, 44. *See also names of specific constellations*
Contact (Sagan), 175–176
Copernicus, Nicolaus, 2, 38, 40, 42, 83–85
Cosmic Background Explorer (COBE), 76, 122–123
Cosmic calendar, 6–7, *6–7*
Cosmic distance scale, 46, 48, 51–52
Cosmic microwave background, 75–76, 121–125, *122*
Cosmic rays, 16, 32
Cosmological constant, 148–149
Cosmological horizon, 69
Critical density, 126, 129, 131, 136, 161–162
Crommelin, Andrew, 118
Curtis, Heber, 61–62

D

Dark clusters, 161
Dark matter
    Big Bang and, 163–164
    composition, 163–164
    definition, 162
    gravitational lensing evidence, 159–163
    gravity evidence, 155–157
    intergalactic space, 157
    neutrinos and, 164–165
    ratio to luminous matter, *157*, 161, 166
    WIMPs, 164–166
    X-ray–emitting gas evidence, 157–159
Darwin (space mission), 97
Days, names of, 80
De docta ignorantia (Nicholas of Cusa), 2
Degeneracy pressure, 111
Democritus, 3–4
Density
    cosmological constant and, 149
    critical, 126, 129, 131, 136, 161–162
    luminous matter, 162
    quantum ripples, 130
    universe, 125–126, 131, 136, 166
Deuterium, 28, 33–34, 164
Disks (galaxy), 65, *65*
Disk stars, 66–67
Distance
    cosmic distance scale, 46, 48, 51–52
    galaxies, 48–52, 55–56, 138–143
    to gamma-ray bursts, 105–106
    lookback times, 68–70
    parallax measurements, 45–46
    Sun, 23–24

Gravitational waves, 109, *109*
Gravity
    containment of elements, 59
    as curvature of spacetime,
        116–119, *117*
    curvature of starlight, 117–118,
        159–160
    discovery of, 115–116
    evidence for dark matter,
        155–157
    expansion of the universe and, 4,
        125–126
    expansion rate and, 141–142
    galaxy clusters, 154–155
    near a neutron star, 103
    star evolution and, 110–111
    structure of the universe and, 56,
        64
The Great Debate, 61–62
Great Galaxy in Andromeda, 3, 5–6,
    60, 60–61, 156, Color Plate 1
Great Wall of galaxies, 53
GUTs (grand unified theories), 132

H

Halo (galaxy), 65, *65,* 164
Halo stars, 66–67
Hawking, Stephen, 150
HD 209458, 92
Heavy water, 33–34. *See also*
    Deuterium
Helioseismology, 31
Helium, 5, 27–28, 123
Helmholtz, Hermann von, 24
Hewlett, Bill, 177
High Energy Transient Explorer
    (HETE), 113
Hipparchus, 82
Hipparcos, 46
Homestake experiment, 26–28, *27*
Hoyle, Fred, 120–121

Hubble, Edwin Powell, 4, 48, *62,*
    62–64
Hubble constant, 140–142, 144,
    147–148
Hubble Deep Field, 64, Color Plate 7
Hubble's law, 48–51, 76
Hubble Space Telescope, 64, 143,
    Color Plate 6
Huchra, John, 52
Human development. *See* Life on Earth
Huxley, Thomas H., 181
Huygens, Christiaan, 183
Huygens (space probe), 20–21, *21*
Hydrogen
    deuterium, 28, 33–34, 164
    evidence for Big Bang, 123
    fusion of, 5, 27–28
Hyperspace flight, 177

I

Inflation, 119, 126–133, *127,* 149
Interferometry, 95–97
Intergalactic space, 158
Intrinsic brightness, 44, 46–48, 63–64,
    145. *See also* Standard candles
Io, *41*
Island universes, 61

J

Jupiter, *41, 82,* 90

K

Kamioka Nucleon Decay Experiment,
    29
Kant, Immanuel, 58, 61
Kelvin, Lord (Thomson, William), 24,
    167
Kelvin scale, 122
Kepler, Johannes, 40–42
Kepler (space mission), 94, 96

Moon (of Earth), 17, 81–82
Moons, 18–21, 87–88
Moore, Gordon, 177
Mount Wilson Observatory, 62–63
Muon neutrinos, 31–33

N

Neutrinos, 23–35
    dark matter and, 164–165
    detection, 26–30, *27, 30,* 33–34
    direction of, 29–30
    formation of, 28, 111
    mass, 25, 35, 165–166
    oscillations between types,
        31–33, 35
    solar neutrino problem, 23–35
    sterile, 34–35
    types, 31–34
Neutron stars, 103–104, *109,*
    109–112, *111*
Newton, Sir Isaac, 84, 115
Newton's first law of motion, 41
Next Generation Space Telescope
    (NGST), 74
Nicholas of Cusa, 2, 38
Nuclear fusion, 5, 25, 27–28, *28,* 72,
    123
Nuclear test ban treaties, 100–101
Nulling, 96

O

Orbits, 66–67, 90, 93, 116, 155–156

P

Packard, Dave, 177
Paine, Thomas, 134
Parallax, 38–39, *39,* 42, 45–46
Parsecs, 141
Pathfinder, 14
51 Pegasi, 89, *91*

Pegasus, *60*
Period–luminosity relation, *63*
Perlmutter, Saul, 147
Phillpotts, Eden, 1
Pickering, Edward, 47
Planck (space mission), 76–77, 131
Planetary motion, 41–42, 81–84, *82,*
    85
Planetesimals, 86
Planets, 79–97
    in binary star systems, 88
    detection, 88–92
    early models, 82–83, *83*
    Earth-like, 85–87, 89–90, 173
    formation of, 86–87, 94
    giant outer, 87–88
    life on, 9, 11–15, 18–21
    meaning of term, 80
    motion, 41–42, 81–84, *82, 85*
    names of days and, 80
    orbital distances and masses, *93*
    telescopes for, 94–95
Plato, 82
Pluto, 88
Pontecorvo, Bruno, 32
Prime Directive, 179
Protogalactic clouds, 66, 68, 74–75
Proxima Centauri, 45
Ptolemy, 82–83, *83*
Pythagoras, 38

Q

Quantum mechanics, 130, 167
Quasars, 70–75, 107
Queloz, Didier, 89

R

Radio waves, 95, 100–101, Color Plate
    11
The Red Limit (Ferris), 122

Thomson, William (Lord Kelvin), 24, 167

Time machines, 68

Titan, 9, 19, 21, *21*

Truth, Sojourner, 14

Type Ia supernovae, 144–147

U

UFOs, 170–172, 177

Ultraviolet waves, 100

Uniform radiation, *108,* 108

Universe

    age of, 4, 6–7, 56, 69, 141–144, 147–148

    brightness of objects in, 44–45

    bubble analogy, 137–140

    as celestial sphere, 38, *44*

    chemical composition of, 123, 154

    closed, 126, *128,* 128–129

    dark matter in, 161

    density of, 125–126, 129, 131, 136, 161–162, 166

    depth perception of, 43–45, *44*

    Earth-centered model, 38–39

    expansion, 4, 48–51, *49,* 120–121, 136–137, *137*

    expansion rate, 135, 138–142, *140,* 147–148

    fate of, 150, 161–162, 166

    flat, 126, *128,* 129, 131

    great debate on scale, 61–62

    island, 61

    large-scale structure, 43–56

    lookback times, 68–70

    mapping, 52–56, *53, 56*

    multiple, 132

    observable, 6, 126, 129

    open, 126, *128,* 126–129, 131, 136

    origin of, 3–4

    postcard from Earth, *3*

    shapes, *128,* 128–129, *137*

    theory of inflation, 119, 126–133, *126,* 149

Upsilon Andromeda, 92

Uranus, 80

V

Vela satellites, 101–102

Venus, 17, 85–86

Verne, Jules, 8

Volcanic vents, 17, 19

W

The War of the Worlds (Wells), 10

Warp drive, 177

Water

    on Callisto and Ganymede, 19

    on Europa, *18,* 18–19

    on Mars, 12–14, *13*

    on Titan, 19–20

Weakly interacting massive particles (WIMPs), 164–166

Wells, H. G., 10

White dwarfs, 110–111, *111,* 145

White-dwarf supernovae, 144–147, *146*

Women astronomers, 47, *47,* 155

X

X-ray bursts, 103–104

X rays, 100–101

X-ray telescopes, 75, 113, 158

Z

Zwicky, Fritz, *154,* 154–155, 159

The universe has been expanding ever since its hot and dense beginning in the Big Bang. Each of the three cubes represents the same region of the universe, showing how the region expands with time.

Our cosmic origins. All the matter and energy in the universe was created in the Big Bang. This sequence of paintings shows the progression of that matter and energy from the Big Bang to human life. Note that the elements from which we are made were produced in stars that shined long ago, and these elements formed the Earth, thanks to the recycling role played by our galaxy.

The Earth was built with elements produced in the stars that lived and died in the Milky Way before our solar system formed.